P9-CKL-815

COUNTER-CURRENT EXTRACTION

COUNTER-CURRENT EXTRACTION

An Introduction to the Design and Operation of Counter-current Extractors

BY

STANLEY HARTLAND

READER IN CHEMICAL ENGINEERING,
UNIVERSITY OF NOTTINGHAM,
ENGLAND

Sometime Scholar of
St. Catharine's College,
Cambridge

PERGAMON PRESS

OXFORD · LONDON · EDINBURGH · NEW YORK
TORONTO · SYDNEY · PARIS · BRAUNSCHWEIG

Pergamon Press Ltd., Headington Hill Hall, Oxford
4 & 5 Fitzroy Square, London W.1
Pergamon Press (Scotland) Ltd., 2 & 3 Teviot Place, Edinburgh 1
Pergamon Press Inc., Maxwell House, Fairview Park, Elmsford, New York 10523
Pergamon of Canada Ltd., 207 Queen's Quay West, Toronto 1
Pergamon Press (Aust.) Pty. Ltd., 19a Boundary Street, Rushcutters Bay, N.S.W. 2011, Australia
Pergamon Press S.A.R.L., 24 rue des Écoles, Paris 5e
Vieweg & Sohn GmbH, Burgplatz 1, Braunschweig

First edition 1970

Library of Congress Catalog Card No. 69-17867

Printed in Germany
08 012976 5

To my mother

ELLEN HARTLAND

Contents

Appendices

Preface

THIS book has grown out of a Ph.D. thesis and has retained its original form to a large extent. The thesis in its turn was initiated by the fact that formulae for calculating numbers of stages and transfer units in Editions III and IV of the *Chemical Engineers' Handbook* persistently gave unreasonable results. Some time elapsed before I became convinced that the error was not in my arithmetic! Closer investigation of the literature revealed a multiplicity of formulae many of which were inconsistent amongst themselves, partly because the useful technique of inversion is not always clearly understood. To introduce some order, all possible formulae were exhaustively examined and this showed that the most convenient equation has been largely overlooked. Its simplicity made it easier to tackle more complex problems involving both forward and back extraction. When the individual flows are counter current, the minimum number of stages or transfer units for a given separation and overall operating conditions was surprisingly given by the identical form of equation. Furthermore, this was also true when the number of cross-current stages were considered.

The optimisation technique used in Chapter 4 was the classical one of setting the derivative equal to zero. There are surprisingly few examples of this available compared with those employing more recent techniques, and so the optimisation procedure has been described in a fair amount of detail. This is also true of the least squares procedures in Chapter 6; most people are familiar with regressions of x on y and vice versa, but not with the determination of the functional relationship which exists between them. The latter is usually of more interest to scientists and engineers, as pairs of variables are generally interdependent, rather than one depending solely on the other. A further statistical element is introduced in Chapter 3, which considers the effect of errors on the calculation of numbers of stages and transfer units. The basic theory relating the error of a function to the errors in the individual variables is well known. However, its importance is such, that it was felt worth while to present a formal treatment as an example of the technique.

Results are presented for both differential and stagewise cases throughout, to illustrate the analogy between the two. A known solution for one case often suggests a method of attack, when solution for the other is difficult. Finite difference techniques have been widely used as they give much simpler and more elegant solutions to stagewise problems than more lengthy proofs by induction. Computers were extensively used to provide the results presented throughout the book and five programs are included. These are intended to be self-explanatory and to illustrate simple computing techniques; in some programs graphical procedures are included. Particular attention has been paid to graphical presentation. It is often possible to achieve a linear plot by careful rearrangement of the equation, and by correct choice of the scale on each axis. Linear equilibrium and operating lines are assumed in order to obtain concise equations wherever possible. Obviously, such an idealised situation does not often occur in practice, but it is useful to have simple and quick answers in preliminary calculations. These may provide a guide as to the best course of action, or a start to a more complicated iterative method. A number of worked examples are included to illustrate the theoretical techniques described.

To keep the main text as brief as possible, frequent use has been made of appendices. These are inserted at the end of the chapters to which they refer. This provides an opportunity of discussing topics in isolation and in more detail than otherwise possible. Lengthy tables are also inserted in appendices. Sections, formulae, tables and figures are numbered on the decimal system based on the chapter in which they occur. For example, Section 2.1, equation 2 (1), Table 2.1 all occur in different parts of Chapter 2. Appendices are also based on the chapter in which they occur, prefixed by the letter A. For example, equation A6b(5) is found in the second appendix of Chapter 6. References are numbered consecutively throughout each chapter and a list is given at the end of each chapter. The notation is consistent throughout the book except for a few local exceptions in appendices associated with computer programs. A general notation list is given at the beginning of the book and subsidiary lists at the end of each chapter.

To summarise, it can be seen that mathematical techniques are emphasised as tools as well as means to an end. One aim of the book is to provide illustrations of these techniques and another is to provide useful results and conclusions pertaining to counter-current extraction. These aims grew out of the initial desire to clarify, unify and systemise the basic equations. It is hoped the book will be useful to undergraduates, research workers and those engaged in the design and operation of counter-current extractors.

I would like to thank Professor R. Edgeworth Johnstone (until recently Head of the Chemical Engineering Department at Nottingham University) for providing the freedom necessary for a member of staff to pursue his research interests. At this stage much sound advice was also provided by the author's colleagues. The subsequent transformation into book form is largely due to the encouragement of Professor P. N. Rowe of University College, London, and Professor P. M. C. Lacey of Exeter University. Most of the results were provided with patience and courtesy by the computing centres at Nottingham University, Manchester University and Harwell. Chapter 6 was written together with Victor Ang whilst the latter was still an undergraduate at Nottingham University. Finally I would like to thank my wife, Ena Margaret, for shielding me from the endless domestic tasks which this book has prevented me undertaking.

The University, Nottingham S. Hartland

Glossary

Absorption. Defined as transfer from G to L phase (in a single contactor).

Back extraction. Transfer from intermediate solvent S to final solvent G (or L).

Cross-current extraction with recirculation. Denotes forward and back extraction in which the intermediate solvent is recirculated between pairs of forward and back extraction stages as shown in Fig. 5.1. The overall flow of the initial and final solvents is still counter-current.

Errors. Refers to variations or uncertainties in the quantities N, J or Q. For example, J may vary due to a change in flowrate or there may be an error in J because of inaccuracy in a flowmeter.

Extraction. Transfer of a solute between phases.

Forward extraction. Transfer from initial solvent L (or G) to intermediate solvent S.

Functional relationship. In general there will be errors in both x and y and one would like to estimate y from x and also x from y. This may be done using the functional relationship which gives the true variation of x with y. It lies between the regression of y on x and the regression of x on y.

Inversion. Refers to physical inversion of the relevant figure. This indicates that symbols should be interchanged as follows so that L, G, S, x_{in}, x_{out}, y_{in}, y_{out}, z_t, and z_b become G, L, S, y_{in}, y_{out}, x_{in}, x_{out}, z_b and z_t respectively.

Optimisation of forward and back extraction. Denotes the objective of minimising $N_1 + N_2$, the total number of stages or overall transfer units (based on the initial and final solvents) for a given separation Q_o and overall operating conditions J_1J_2. It involves determining the optimum values of N_1/N_2 and the individual extraction factors J_1 and J_2.

An equivalent alternative is the objective of maximising Q_o for a given $N_1 + N_2$ and J_1J_2. This also involves determining the optimum values of N_1/N_2, J_1 and J_2.

Note that optimisation of the stagewise case includes the possibility of cross-current extraction.

Regression. The regression of y upon x estimates the value of y for a given value of x, assuming there is no error in x.

The regression of x upon y estimates the value of x for a given value of y, assuming there is no error in y.

Stripping. Defined as transfer from L to G phase (in a single contactor).

Notation

a	interfacial area per unit volume
A	total area of contact between two phases
$\mathscr{A}, \mathscr{B}$	integration constants
b	defined by equation 6(12)
c	intercept of equilibrium line on y-axis
C	specific heat
d	density
D	sum of squares of deviations
e	standard deviation
e^2	variance
E	overall stage efficiency
E_m	Murphree stage efficiency based on concentration $y\,(E_m = t/(1 + t))$
G	flowrate of phase in which solute concentration is y
h	distance along differential contactor
H	total length of a differential contactor
Htu	height of overall transfer unit
$\mathscr{H}$	Henry's Law constant
J	extraction factor ($J = mG/L$)
J^*	inverse extraction factor ($J^* = L/mG$)
k^2	ratio of variance in y to variance in x (in functional relationship)
K	overall extraction factor in forward and back extraction ($K = J_1J_2 = m_1m_2G/L$)
k_g	individual mass transfer coefficient for the phase G
k_l	individual mass transfer coefficient for the phase L
k_G	overall mass transfer coefficient based on phase G
k_L	overall mass transfer coefficient based on phase L
L	flowrate of phase in which solute concentration is x
m	slope of equilibrium line (equation $y = mx + c$)
M	slope of equilibrium line through origin (equation $y = Mx$)
n	number of typical stage

N	number of theoretical stages or transfer units
N_A	number of actual (non-equilibrium) stages
N_G	number of overall transfer units based on phase of flowrate G
N_L	number of overall transfer units based on phase of flowrate L
N_S	number of theoretical stages
N_T	number of transfer units
p	number of pairs of observations
P	total number of stages or transfer units ($P = N_1 + N_2$)
Q	separation factor (defined by Table 2.3)
Q^*	inverse separation factor (defined by Table 2.3)
Q_o	overall separation factor in counter-current forward and back extraction
s	cross-sectional area of contactor
S	flowrate of intermediate solvent (in which solute concentration is z)
t	stagewise transfer coefficient based on phase G ($t = k_G aV/G$)
u	superficial velocity of L phase
U	overall heat transfer coefficient
v	superficial velocity of G phase
V	volume of a stage
W	heat capacity of stream per unit time
x	solute concentration in phase of flowrate L
$\bar{x}$	arithmetic mean x concentration
$x(y)$	x concentration in equilibrium with concentration y ($x(y) = (y - c)/m$ for a single contactor $x(y) = y/m_1m_2 - c_1/m_2 - c_2/m_1m_2$ for forward and back extraction)
$x(z)$	x concentration in equilibrium with concentration z ($x(z) = y/m_1 - c_1/m_1$ for contactor 1)
X	best estimate of x for a given value of y
y	solute concentration in phase of flowrate G
$\bar{y}$	arithmetic mean y concentration
$y(x)$	y concentration in equilibrium with concentration x ($y(x) = mx + c$ for a single contactor $y(x) = m_1m_2 + m_2c_1 + c_2$ for forward and back extraction)
$y(z)$	y concentration in equilibrium with concentration z ($y(z) = m_2z + c_2$ for contactor 2)

Δy	concentration difference based on y
Δy_{lm}	log mean concentration difference based on y
Y	best estimate of y for a given value of x
z	solute concentration in intermediate solvent of flowrate S
$z(x)$	z concentration in equilibrium with concentration x ($z(x) = m_1 x + c_1$ for contactor 1)
$z(y)$	z concentration in equilibrium with concentration y ($z(y) = y/m_2 - c_2/m_2$ for contactor 2)

Greek symbols

α	extraction factor for non-equilibrium stages ($\alpha = (1 + Jt)/(1 + t)$)
β	overall extraction factor in cross-current extraction with recirculation ($\beta = J_2(1 + J_1)/(1 + J_2)$)
δ	a small increment
Δ	a finite difference
θ	temperature
$\Delta\theta_{lm}$	log mean temperature difference
λ	overall extraction factor in cross-current extraction ($\lambda = J/(1 + J)$)
μ	index of $J_1 J_2$ (in approximate solution for existing forward and back stagewise extractor)
ξ	dimensionless x concentration based on concentrations x_{in} and y_{in} defined by equations 1(12) and 1(18)
Σ	indicates summation over total number of observations
ϕ	dimensionless y concentration based on concentrations x_{in} and y_{in} defined by equations 1(13) and 1(19)
χ	dimensionless x concentration based on concentrations x_{in} and x_{out} defined by equations 1(15) and 1(21)
ψ	dimensionless y concentration based on concentrations y_{in} and y_{out} defined by equations 1(16) and 1(22)

Subscripts

in	refers to stream entering contactor
out	refers to stream leaving contactor

A	refers to actual stages
b	refers to intermediate solvent leaving contactor 2
C	refers to cross-current extraction
G	refers to phase of flowrate G
i	refers to ith pair of observations
L	refers to phase of flowrate L
n	refers to nth stage
$n - 1$	refers to $(n - 1)$th stage
$n + 1$	refers to $(n + 1)$th stage
o	refers to overall separation in counter-current forward and back extraction
O	refers to optimum conditions in counter-current forward and back extraction
p	refers to pth pair of observations
R	refers to cross-current extraction with recirculation
t	refers to intermediate solvent leaving contactor 1
x	refers to regression of x on y
y	refers to regression of y on x
1	refers to contactor 1
2	refers to contactor 2
1, 2 ···	refers to 1st, 2nd ··· pairs of observations

Superscript

*	denotes inversion

The above quantities may be expressed in any consistent set of units in which force and mass are not defined independently.

CHAPTER 1

Introduction

THIS book considers several aspects of the transfer of a single solute between two immiscible phases in counter-current flow and the conclusions apply to gas–liquid and liquid–liquid transfer. For the case of straight operating and equilibrium lines the number of stages or transfer units, N, may then be expressed in terms of an extraction factor J, and a separation factor Q. Although these idealised conditions do not usually occur in practice they enable a quick and relatively easy answer to be obtained which is often satisfactory for preliminary design. The results may provide a reasonable start to a more precise and complicated method thus limiting the more tedious work involved or saving computer time. Concise analytical solutions often provide a feel for the problem which is difficult to obtain otherwise. Even in real systems, to which they do not quantitatively apply, they indicate the qualitative effect of the variables involved.

1.1. Basic Equations

A contactor is shown in Fig. 1.1 for which the counter-current flows of the two immiscible phases are L and G, and the end concentrations are x_{in}, y_{out} and x_{out}, y_{in}. The equation of the equilibrium line of slope m and intercept c is taken as

$$y = mx + c \qquad 1(1)$$

where y is the concentration in phase G and x in phase L.

An extraction factor may then be defined as

$$J = \frac{mG}{L} \qquad 1(2)$$

and a separation factor as

$$Q = \frac{y_{\text{in}} - y(x_{\text{in}})}{y_{\text{out}} - y(x_{\text{in}})} \qquad 1(3)$$

where $y(x_{\text{in}})$ is the y concentration in equilibrium with x_{in} so that

$$y(x_{\text{in}}) = mx_{\text{in}} + c$$

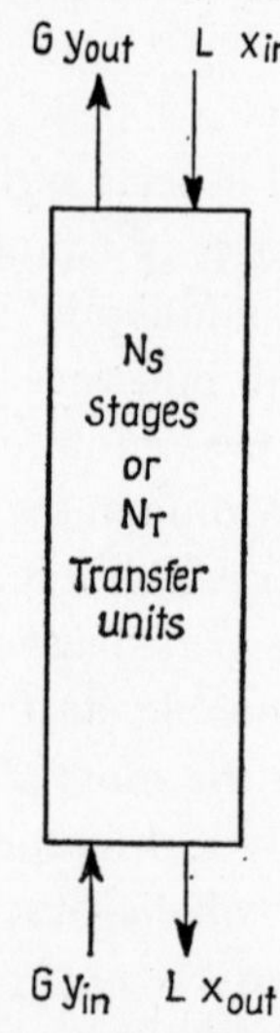

FIG. 1.1. Flowrates and end concentrations for stagewise or differential contactor.

Then for a process having N_S stages

$$N_S = \frac{\ln [(1 - J)\,Q + J]}{\ln 1/J} \qquad 1(4)$$

and for a differential process having N_G overall mass transfer units based on the G phase

$$N_G = \frac{k_G A}{G} = \frac{\ln [(1 - J)\,Q + J]}{1 - J} \qquad 1(5)$$

where k_G is the overall mass transfer coefficient based on the G phase and A is the total interfacial area.

These formulae were obtained in the first instance by Kremser[1] and Colburn[2] respectively for the case of absorption with $x_{\text{in}} = 0$.

1.2. Inverse Separation Factors

The separation factor Q only includes three concentrations; by choosing different boundary conditions, x_{out} could have been included in the separation factor and a new relationship between N, J and Q obtained. One such result is

$$Q^* = \frac{x(y_{in}) - x_{in}}{x(y_{in}) - x_{out}} \tag{1(6)}$$

where $x(y_{in})$ is the x concentration in equilibrium with y_{in}.

Q^* is related to Q by

$$Q = \frac{Q^*}{(1 - J^*)\,Q^* + J^*} \tag{1(7)}$$

where

$$J^* = \frac{L}{mG} = \frac{1}{J} \tag{1(8)}$$

which may be verified using the overall balance

$$y(x_{in}) - y(x_{out}) = J\,(y_{out} - y_{in}) \tag{1(9)}$$

Substitution of equation 1(7) into equations 1(4) and 1(5) gives

$$N_S = \frac{\ln\,[(1 - J^*)\,Q^* + J^*]}{\ln 1/J^*} \tag{1(10)}$$

and

$$N_L = \frac{k_L A}{L} = \frac{N_G}{J^*} = \frac{\ln\,[(1 - J^*)\,Q^* + J^*]}{1 - J^*} \tag{1(11)}$$

where N_L is the number of overall mass transfer units and k_L is the overall mass transfer coefficient, both based on the L phase. Equations 1(10) and 1(11) are identical in form with equations 1(4) and 1(5) and J^* and Q^* are said to be the inverses of J and Q.

1.3. Graphical Representation

Equations 1(4) and 1(5) are plotted in Figs. 1.2 and 1.3. A logarithmic scale is used for N to improve the gradation of the parameter J when it is less than 1. The asymptotic nature and uneven gradations of the curves for J greater than 1 makes them difficult to use in this region and so it is then better to replace J by J^* (which will be less than 1) and Q by the inverse separation factor Q^* defined by equation 1(6). For example, Colburn[2] states that the economic range of J is 0·5–0·8

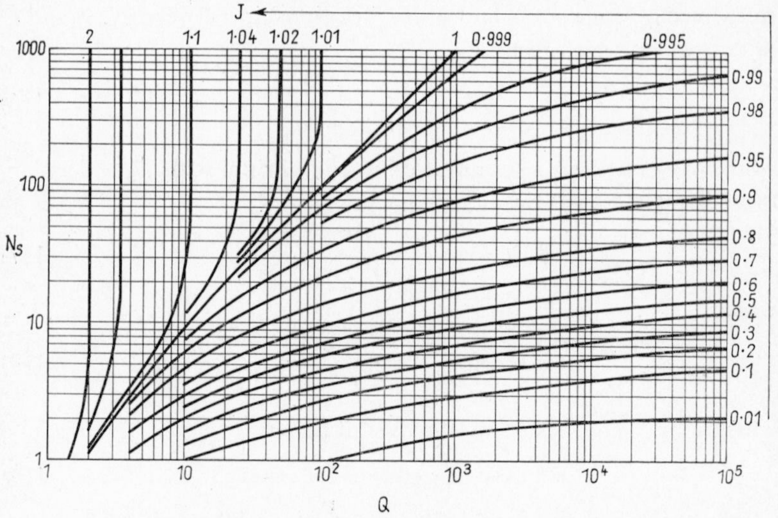

FIG. 1.2. Variation of N_S with Q for different values of J.

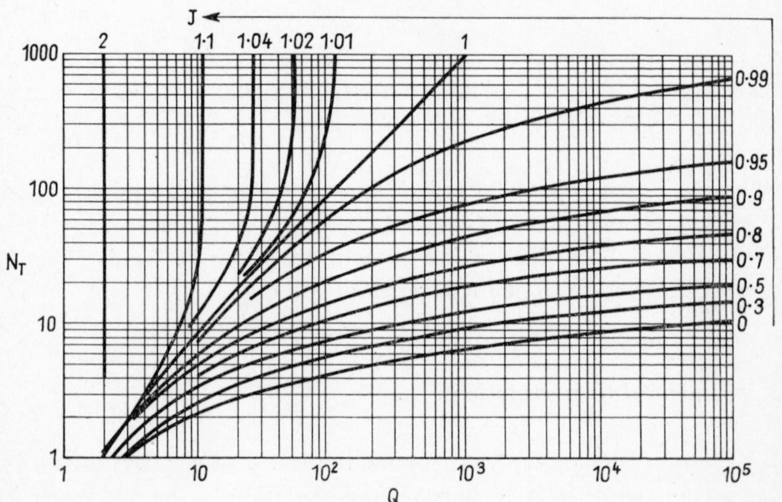

FIG. 1.3. Variation of N_T with Q for different values of J.

for absorption and 1·5–2 for stripping. Reference to Figs. 1.2 and 1.3 shows that this range of J greater than 1 limits Q to between 2 and 3 and even if J drops to 1·1, Q cannot exceed 11. Consideration of the logarithmic arguments in equations 1 (4) and 1 (5) shows that in general the limiting value of Q is given by $J/(J-1)$.

1.4. Absorption and Stripping

Some authors (for example Perry[(3)] and Sherwood and Pigford[(4)]) suggest the use of Q and J for absorption and Q^* and J^* for stripping. This involves the use of J and J^* greater than 1 in certain cases which

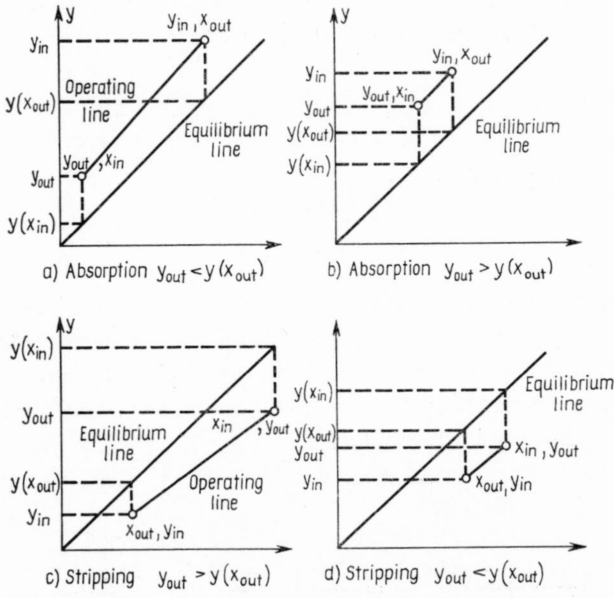

FIG. 1.4. Relative positions of end concentrations in absorption and stripping.

would lead to inaccurate results. Suppose we define transfer from the G phase to the L phase as absorption and from the L to the G phase as stripping as shown in Fig. 1.4. In absorption y_{in} is greater than y_{out}

and x_{out} is greater than x_{in}, in stripping x_{in} is greater than x_{out} and y_{out} is greater than y_{in}. If the aim in absorption is to achieve a low value of y_{out} then J is less than 1 but if the aim is to achieve a high value of x_{out} then J is greater than 1. In stripping, a high value of y_{out} corresponds to J less than 1 and a low value of x_{out} co responds to J greater than 1. It is thus best not to think in terms of absorption or stripping but to consider whether J is greater or less than unity.

1.5. Concentration Profiles

Stagewise

As shown in Appendix 1a, the variations in the concentrations x and y with the numbers of stages n are given by

$$\xi_{n+1} = \frac{y(x_{n+1}) - y(x_{in})}{y_{in} - y(x_{in})} = \frac{J^N - J^n}{J^N - 1/J} \qquad 1(12)$$

and

$$\phi_n = \frac{J(y_{in} - y_n)}{y_{in} - y(x_{in})} = \frac{J^n - 1}{J^N - 1/J} \qquad 1(13)$$

The profiles are expressed in terms of x_{in} and y_{in} as the inlet concentrations are more likely to be known than the outlet concentrations. Note that the dimensionless concentrations ξ_{n+1} and ϕ_n are based on x_{n+1} and y_n, respectively, as these are the concentrations of the streams passing each other in the contactor.

When $n = 0$, $x_1 = x_{out}$ and when $n = N$, $y_N = y_{out}$ so equations 1(12) and 1(13) become

$$\frac{y(x_{out}) - y(x_{in})}{y_{in} - y(x_{in})} = \frac{J(y_{in} - y_{out})}{y_{in} - y(x_{in})} = \frac{J^N - 1}{J^N - 1/J} \qquad 1(14)$$

which were derived by Souders and Brown.[5] These overall separation factors are directly related by the overall balance 1(9) and are logical in the sense that they represent the ratio of the actual concentration difference to the maximum possible concentration difference.

Dividing equations 1(12) and 1(13) in turn by 1(14) gives

$$\chi_{n+1} = \frac{y(x_{n+1}) - y(x_{\text{in}})}{y(x_{\text{out}}) - y(x_{\text{in}})} = \frac{J^N - J^n}{J^N - 1} \qquad 1(15)$$

and

$$\psi_n = \frac{y_{\text{in}} - y_n}{y_{\text{in}} - y_{\text{out}}} = \frac{J^n - 1}{J^N - 1} \qquad 1(16)$$

which are also obtained from first principles in Appendix 1a. Adding the two equations gives

$$\chi_{n+1} + \psi_n = 1 \qquad 1(17)$$

The dimensionless concentrations ψ_n and χ_{n+1} are plotted as functions of n/N, at different values of N, for $J = 0{\cdot}7$ in Fig. 1.5.

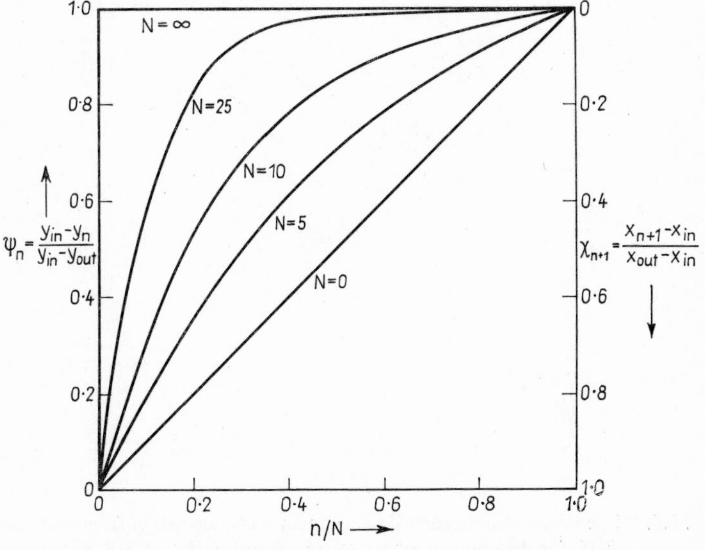

FIG. 1.5. Dimensionless concentrations ψ_n and χ_{n+1} as functions of stage number n for different number of stages N at $J = 0{\cdot}7$.

Differential

For the differential case the concentration profiles, derived in Appendix 1b, are

$$\xi = \frac{y(x) - y(x_{\text{in}})}{y_{\text{in}} - y(x_{\text{in}})} = \frac{e^{N_L(1-1/J)} - e^{N_L(1-1/J)h/H}}{e^{N_L(1-1/J)} - 1/J} \qquad 1(18)$$

and

$$\phi = \frac{J(y(x_{in}) - y)}{y_{in} - y(x_{in})} = \frac{e^{-N_G(1-J)h/H} - 1}{e^{-N_G(1-J)} - 1/J} \quad 1(19)$$

When $h = 0$, $x = x_{out}$, and when $h = H$, $y = y_{out}$ so remembering that $N_L = JN_G$, equations 1(18) and 1(19) become

$$\frac{y(x_{out}) - y(x_{in})}{y_{in} - y(x_{in})} = \frac{J(y_{in} - y_{out})}{y_{in} - y(x_{in})} = \frac{e^{-N_G(1-J)} - 1}{e^{-N_G(1-J)} - 1/J} \quad 1(20)$$

which are the differential expressions corresponding to equation 1(14).

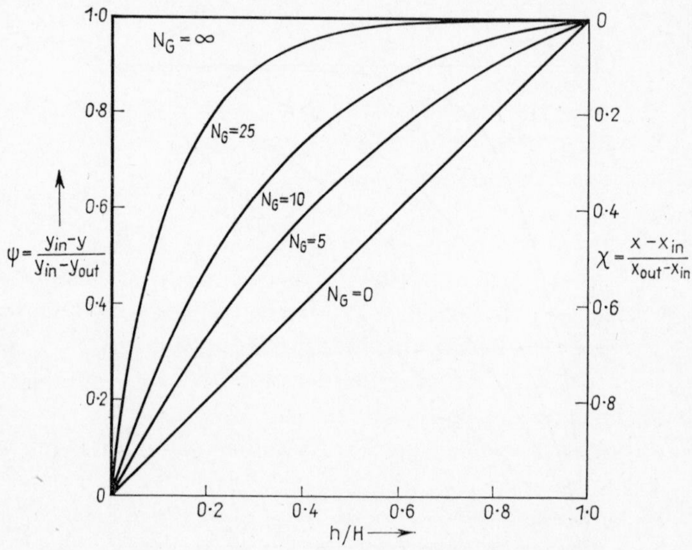

FIG. 1.6. Dimensionless concentrations ψ and χ as functions of fractional distance h/H, for different numbers of transfer units N_G at $J = 0{\cdot}7$.

Dividing equations 1(18) and 1(19) in turn by equation 1(20) gives

$$\chi = \frac{y(x) - y(x_{in})}{y(x_{out}) - y(x_{in})} = \frac{e^{N_L(1-1/J)} - e^{N_L(1-1/J)h/H}}{e^{N_L(1-1/J)} - 1} \quad 1(21)$$

and

$$\psi = \frac{y_{in} - y}{y_{in} - y_{out}} = \frac{e^{-N_G(1-J)h/H} - 1}{e^{-N_G(1-J)} - 1} \quad 1(22)$$

which are also obtained from first principles in Appendix 1b. Adding the two equations gives

$$\chi + \psi = 1 \qquad 1(23)$$

The dimensionless concentrations χ and ψ are plotted as functions of h/H for different values of N_G ($= N_L/J$) at $J = 0{\cdot}7$ in Fig. 1.6.

1.6. Application of the Equations

As stated earlier, equations 1(4), 1(5), 1(10), 1(11), 1(14) and 1(20) apply to the transfer of a single solute between two immiscible phases in counter-current flow when there is a linear equilibrium relationship. Isothermal conditions of operation are implied and this may not always be so, especially in the case of gas absorption.

For the operating line to be straight the ratio of the flows (which is the slope of the operating line) must be constant. This will usually be so when the concentrations are expressed in mass ratios for the flow of each solvent alone is always constant when the two are immiscible (even though the flow of solution as a whole may not be). Flows may also be expressed per unit cross-sectional area of the contactor, s without any loss of generality. This introduces new variables $u = L/s$ and $v = G/s$ which are the velocities of the L and G phases.

The derivation of the equation for the differential case (as, for example, in Appendix 1b) is such that they only apply when the interfacial area per unit volume, a, and the overall mass transfer coefficients, k_G or k_L, are constant. The latter will be so when the individual mass transfer coefficients, k_g and k_l, for the two phases G and L are constant, for the coefficients are related by:

$$\frac{1}{k_G} = \frac{1}{k_g} + \frac{m}{k_l} \qquad 1(24)$$

and

$$\frac{1}{k_L} = \frac{1}{k_l} + \frac{1}{mk_g} \qquad 1(25)$$

so that

$$k_L = mk_G$$

There is no such restriction on the stagewise equations (derived, for example, in Appendix 1a) as it is assumed that the stages are at equilibrium which implies $k_G a$ and $k_L a$ tend to infinity. When the stages are not at equilibrium it is usual to allow for this by dividing the number of theoretical stages N_S by the overall stage efficiency E to obtain the number of actual stages N_A. More accurately, individual stage efficiencies may be used as described in Section 8.9.

1.7. Units and Dimensions

The dimensionless variables J and Q contain the fundamental variables L, G, x, y, m and c. These must be expressed in consistent units and some typical units and dimensions are given in Table 1.1 In this table M is mass, V is volume and T is time. Typical units of V are m^3 or ft^3

TABLE 1.1. UNITS AND DIMENSIONS OF FUNDAMENTAL VARIABLES

Variable	L and G	x, y and c	m
Dimensions	$\frac{M}{T}$	$\frac{M}{M}$	$\frac{M}{M} \Big/ \frac{M}{M}$
Typical units	kilograms of solution / second	kilograms of solute / kilograms of solution	$\frac{y}{x}$
	lb moles of solvent / hour	lb moles of solute / lb moles of solvent	$\frac{y}{x}$
Dimensions	$\frac{V}{T}$	$\frac{M}{V}$	$\frac{M}{V} \Big/ \frac{M}{V}$
Typical units	m^3 of solution / minute	gram moles of solvent / m^3 of solution	$\frac{y}{x}$
	gallons of solvent / day	lb of solute / gallons of solvent	$\frac{y}{x}$

and of T seconds or hours. M may be in kilograms, gram moles, lb or lb moles; and may refer to the solution or solvent. m is dimensionless but its value depends on the dimensions of x and y, as interconversion of concentrations involves the densities of the L and G phases which will usually be different. If the density of the solution varies with concentration of the solute the equilibrium line may be straighter in one set of dimensions than in another.

1.8. Straight Equilibrium Lines

One of the major assumptions throughout this book is that of linear equilibrium lines so it is important to consider when these will occur.

Many equilibrium lines are straight when the solutions are dilute. As the equilibrium line usually passes through the origin c is zero in these cases. However, inclusion of the intercept c extends the usefulness of the work as some equilibrium lines have straight portions which do not pass through the origin. The distribution of uranyl nitrate (UN) between water and methyl isobutyl ketone (MIBK) is one such example.[6] When the concentration x, of UN in the MIBK is between 0·09 and 0·35 expressed as a weight ratio, the equilibrium follows the relationship, $y = 0{\cdot}66x + 0{\cdot}24$, almost exactly. In addition curved equilibrium lines may be approximated to straight lines over parts of their length. For example, the curve $y = \sqrt{x}$ may be approximated in the region $x = 1{\cdot}0$ to $2{\cdot}9$ by $y = 0{\cdot}337x + 0{\cdot}64$ to an accuracy of better than $\pm 2\%$. This is comparable with that of experimental equilibrium data which has been assumed to be linear.[7]

1.9. Comparison with Cross-current Extraction

The stagewise extraction of a single solute from a solvent of flowrate G with a cross-current solvent of flowrate L per stage when there is a linear equilibrium relationship

$$y = mx + c$$

is considered. As shown in Appendix 1c, the number of stages N_C required to give a separation Q_C is given by

$$\lambda = \frac{J}{1 + J}$$

where

$$N_C \ln 1/\lambda = \ln Q_C \qquad 1(26)$$

and

$$Q_C = \frac{y_{\text{in}} - y(x_{\text{in}})}{y_{\text{out}} - y(x_{\text{in}})}$$

The solute concentrations y_{in} and y_{out} refer to the feed and product of the solvent G and the concentration x_{in} to the feed of the cross-current solvent L.

The separation factor Q_C is of the form Q defined by equation 1(3). Counter-current extraction is described by equation 1(4),

$$N_S \ln 1/J = \ln [(1 - J) Q + J] \qquad 1(27)$$

and is compared with cross-current extraction for the same value of the extraction factor J in Figs. 1.7 and 1.8. The former is a plot of N_C/N_S for a given separation $Q = Q_C$ and the latter a plot of Q/Q_C for a given number of stages $N_S = N_C$.

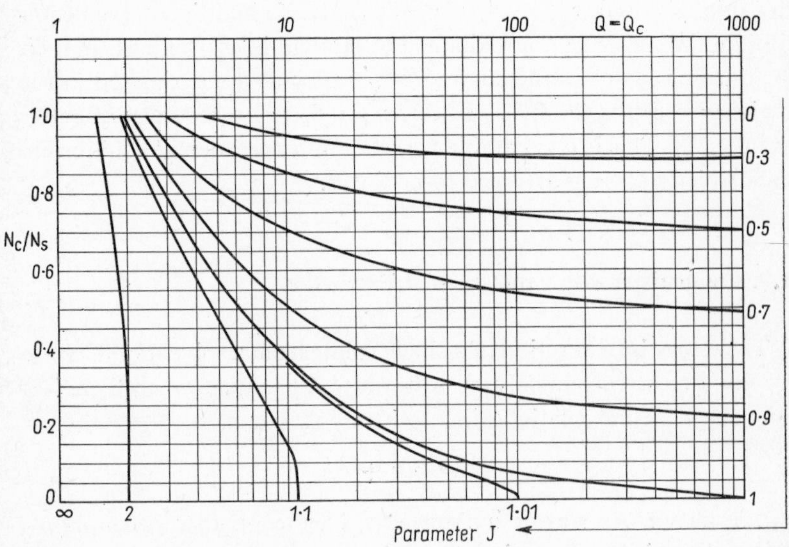

FIG. 1.7. Comparison of counter-current and cross-current extraction. Variation of ratio of numbers of stages N_C/N_S with separation $Q = Q_C$ for different values of extraction factor J.

The ratio of separation factors when $N_S = N_C = N$ is given by

$$\frac{Q}{Q_C} = \frac{(1/J)^N - J}{1 - J} \Big/ \left(\frac{1}{J} + 1\right)^N \qquad 1(28)$$

Writing $J = 1/J^*$ and rearranging this may be rewritten

$$\frac{Q}{Q_C} = \frac{(1/J^*)^N - J^*}{1 - J^*} \bigg/ \left(\frac{1}{J^*} + 1\right)^N \qquad 1(29)$$

which is identical in form with equation 1(28). It thus follows that when J is greater than one it may be replaced by $1/J$ and so only values of J less than one are shown in Fig. 1.8. However, forms of the ex-

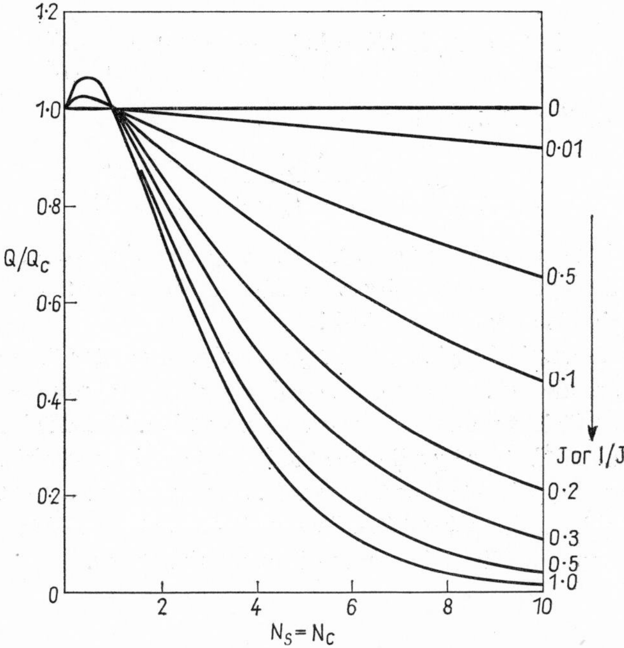

FIG. 1.8. Comparison of counter-current and cross-current extraction. Variation of ratio of separations Q/Q_C with number of stages $N_S = N_C$ for different values of extraction factor J.

pression for the ratio N_C/N_S when $Q = Q_C$ are not identical when written in terms of J and J^*, and so additional values of J greater than 1 are shown in Fig. 1.7.

Both the ratios Q/Q_C and N_C/N_S are less than one when the number of stages is greater than one, and equal to unity when $N_S = N_C = 1$. When $N_S = 1$ it follows from equation 1(27) that $Q = (1 + J)/J = 1/\lambda$

and so equation 1(26) shows that $N_C = 1$ when $Q = Q_C$. In Fig. 1.7 the ratio N_C/N_S (at $Q = Q_C$) may be replaced by $\log Q / \log Q_C$ (at $N_S = N_C$). When $Q = Q_C$ it follows from equation 1(26) that $N_C = \ln Q/\ln 1/\lambda$ and so

$$\frac{N_C}{N_S} = \frac{\ln Q}{\ln [(1 - J) Q + J]} \frac{\ln 1/J}{\ln 1/\lambda} \qquad 1(30)$$

Similarly when $N_S = N_C$,

$$\ln Q_C = N_S \ln 1/\lambda = \ln [(1 - J) Q + J] \frac{\ln 1/\lambda}{\ln 1/J}$$

from equation 1(27) and so

$$\frac{\ln Q}{\ln Q_C} = \frac{\ln Q}{\ln [(1 - J) Q + J]} \frac{\ln 1/J}{\ln 1/\lambda} \qquad 1(31)$$

which is identical with the ratio given by equation 1(30). All three ratios are greater than unity when N_S and N_C are less than unity but this region is only of theoretical interest.

In the practical region when N_S and N_C are greater than one the ratios are always less than one showing that cross-current extraction is more efficient than counter-current extraction. The efficiency may be increased further by adjusting the flowrate and composition of the cross-current solvent along the cascade. However, cross-current extraction requires more solvent (the flowrate is L per stage) and the contactor is more complicated because of the multiple feeds.

1.10. A General Survey

Many different separation factors are possible in addition to those defined by equations 1(3), 1(6), 1(14) and 1(20) and many relations between N, J and Q have appeared in the literature, some of them conflicting. The next chapter resolves this conflict by considering all possible separation factors and deriving the relationships between N, J and Q for both the differential and stagewise cases. There are fourteen basic separation factors but because the x and y phases may be interchanged there are only eight basic formulae for N in each of which J

and Q may be replaced by their inverses. All the formulae may be used in heat transfer calculations if analogous quantities are used.

The most convenient formula for general use in which the separation factor is the reciprocal of its inverse is

$$N_S \ln 1/J = N_T (1 - J) = \ln Q \qquad 1(32)$$

where

$$J = \frac{mG}{L}, \quad Q = \frac{y_{\text{in}} - y(x_{\text{out}})}{y_{\text{out}} - y(x_{\text{in}})} \quad \text{and} \quad N_T = N_G$$

or

$$J = \frac{L}{mG}, \quad Q = \frac{y_{\text{out}} - y(x_{\text{in}})}{y_{\text{in}} - y(x_{\text{out}})} \quad \text{and} \quad N_T = N_L$$

This formula may be used to calculate N, J and Q from the other two variables. Experience shows that such calculations often lead to large errors (the effects of a "pinch", for example, are well known). Chapter 3 considers the interrelation of variations in N, J and Q. It shows that the error in N resulting from an error in J can be considerable but that errors in Q usually have a small effect. However, the error in Q itself may be large so both must be taken into account. Errors in J and N can give rise to large errors in Q. The derivation of errors in J and Q from the experimental measurements of flowrates, end concentrations and equilibrium data is discussed.

In Chapters 4 and 5 the optimisation of forward and back extraction is considered. When it is necessary to separate two solutes, a solvent having high selectivity and extracting power for one of them must be used. If this solvent is to be recirculated the solute must be back extracted into a further solvent which is more convenient for use in the next stage of the process.

Forward and back extraction thus involves the transfer of a single solute from an initial to an intermediate solvent in one contactor and thence to a final solvent in a second contactor, with recycle of the intermediate solvent. For a given separation Q_O and overall operating conditions $J_1 J_2$ $(= m_1 m_2 G/L)$ the minimum number of theoretical stages $N_{S1} + N_{S2}$ or overall transfer units based on the initial and final solvents $N_{L1} + N_{G2}$ occurs when $N_{L1}/J_1 = N_{G2} = N_{TO}$ or for practical purposes when $N_{S1} = N_{S2} = N_{SO}$. In addition for the differential case, and for the stagewise case when N_{SO} is large, N_O is given by

$$N_{TO}(1 - J_O) = N_{SO} \ln 1/J_O = \ln Q_O \qquad 1(33)$$

where

$$Q_O = \frac{y_{\text{in}} - y(x_{\text{out}})}{y_{\text{out}} - y(x_{\text{in}})}, \quad J_0 = J_1 = J_2 = \sqrt{(J_1 J_2)}$$

$$J_1 = m_1 S/L \quad \text{and} \quad J_2 = m_2 G/S$$

The stagewise approximation is suggested by the differential result. For the stagewise case the optimum values of N_{SO}, J_1, J_2 and Q_O are presented graphically for all N_{SO}.

Under the same conditions of overall extraction (so that $J = J_1 J_2$) ordinary counter-current extraction is always more efficient than forward and back extraction (so that $N < N_O$ for a given separation $Q = Q_O$, and $Q_O < Q$ for a given number of stages or transfer units $N = N_O$).

For an existing differential contactor with fixed overall operating conditions $J_1 J_2$ the maximum separation Q_O is given exactly by

$$\left(\frac{J_1 J_2}{N_1} + \frac{1}{N_2}\right) \ln Q_O = 1 - J_1 J_2$$

and hence the optimum recirculation rate S (in terms of J_1) by

$$\left(1 - \frac{\ln Q_O}{N_1/J_1}\right)\left(1 - \frac{\ln Q_O}{N_2}\right) = J_1 J_2$$

The corresponding approximate expressions for an existing stagewise contactor suggested by the differential expressions are

$$\left(\frac{1}{N_1} + \frac{1}{N_2}\right) \ln Q_O = \ln 1/J_1 J_2$$

and

$$\left(1 - \frac{(J_1 J_2)^{\mu - 1} \ln Q_O}{N_1/J_1}\right)\left(1 - \frac{(J_1 J_2)^{\mu} \ln Q_O}{N_2}\right) = 1 - (J_1 J_2)^{\mu} \ln 1/J_1 J_2$$

where

$$\mu = 0{\cdot}75 \left(\frac{\ln Q_O}{N_2 \ln 1/J_1 J_2}\right)^2$$

These approximations became more accurate as J_1J_2 approaches unity and Q_O becomes large. The exact solution is presented graphically in Appendix 41.

Another way of carrying out forward and back extraction in a stagewise process is by recirculating the intermediate solvent between pairs of stages in contactors 1 and 2. This will be called cross-current extraction with recirculation. It has the advantage that the composition and flowrate of the intermediate solvent are under direct control in each pair of stages and can be varied along the contactor. The number of pairs of theoretical stages N_R required to give a separation Q_R with extraction factors J_1 and J_2 is given by

$$N_R \ln 1/\beta = \ln Q_R \qquad 1(34)$$

where

$$Q_R = \frac{y_{\text{in}} - y(x_{\text{out}})}{y_{\text{out}} - y(x_{\text{in}})}, \quad \beta = \frac{J_2 (1 + J_1)}{1 + J_2}$$

$$J_1 = m_1 S/L \quad \text{and} \quad J_2 = m_2 G/S$$

Under the same conditions of overall extraction, (J_1J_2) cross-current forward and back extraction is always more efficient than counter-current forward and back extraction. Even at the optimum operating condition for counter-current extraction, $N_R < N_{SO}$ for a given separation $Q_O = Q_R$, and $Q_O < Q_R$ for a given $N_R = N_{SO}$.

However, under the same condition of overall extraction $(J = J_1J_2)$ ordinary counter-current extraction is more efficient than cross-current forward and back extraction, (so that $N_S < N_R$ for a given separation $Q = Q_R$, and $Q_R < Q$ for a given $N_S = N_R$).

It will be noted that the form of equations 1(32), 1(33) and 1(34) for differential and stagewise contactors is identical. In other words, the same equation may be used to represent counter-current extraction in a single contactor and the optimisation of forward and back extraction using either counter-current flow or cross-current flow with recirculation.

The above discussion shows how knowledge of a linear equilibrium relationship very much eases mathematical computations in the analysis of extraction problems. In ternary liquid–liquid systems the distribution curve representing the equilibrium of a solute between conjugate phases is usually non linear, but it is often possible to represent the upper and lower concentration regions of the curve by straight lines.

In Chapter 6 a least-squares method is used to investigate the solute distribution in ninety-nine aqueous–organic systems, forty-seven organic–organic systems and thirty-nine liquid–gas systems. The slope and its error is tabulated for each system both for lines through the origin and through the mean values of the observations. In 38% of the aqueous–organic systems, 26% of the organic–organic systems and 77% of the liquid–gas systems there exists a linear region with an error in the slope of less than 1%. When the full concentrations range is considered the corresponding figures are 16%, 2% and 31%.

Finally Chapter 7 reviews the conclusions which have been reached throughout the book. Using these foundations Chapter 8 suggests how the work may be extended to more complicated cases and applied to other fields. Some of the material in the book is summarised in references 8–13.

Notation for Chapter 1

a	interfacial area per unit volume
A	total area of contact between phases L and G
$\mathscr{A}, \mathscr{B}$	integration constants
c	intercept of equilibrium line on y-axis
E	overall stage efficiency
G	flowrate of phase in which solute concentration is y
h	distance along differential contactor
H	total length of a differential contactor
Htu	height of overall transfer unit
J	extraction factor ($J = mG/L$)
J^*	inverse extraction factor ($J^* = L/mG$)
k_g	individual mass transfer coefficient for phase G
k_l	individual mass transfer coefficient for phase L
k_G	overall mass transfer coefficient based on G phase
k_L	overall mass transfer coefficient based on L phase
L	flowrate of phase in which solute concentration is x
m	slope of equilibrium line (equation $y = mx + c$)
n	number of a typical stage

N	number of theoretical stages or transfer units
N_A	number of actual stages
N_G	number of overall transfer units based on phase of flowrate G
N_L	number of overall transfer units based on phase of flowrate L
N_S	number of theoretical stages
N_T	number of transfer units
Q	separation factor (defined by equation 1(3))
Q^*	inverse separation factor (defined by equation 1(6))
s	cross–sectional area of contactor
S	flowrate of intermediate solvent
u	superficial velocity of L phase
v	superficial velocity of G phase
x	solute concentration in phase of flowrate L
$x(y)$	x concentration in equilibrium with concentration y ($x(y) = (y - c)/m$)
y	Solute concentration in phase of flowrate G
$y(x)$	y concentration in equilibrium with concentration x ($y(x) = mx + c$)

Greek symbols

λ	overall extraction factor in cross-current extraction ($\lambda = J/(1 + J)$)
ξ	dimensionless x concentration based on concentrations x_{in} and y_{in}
ϕ	dimensionless y concentration based on concentrations x_{in} and y_{in}
χ	dimensionless x concentration based on concentrations x_{in} and x_{out}
ψ	dimensionless y concentration based on concentrations y_{in} and y_{out}

Subscripts

in	refers to inlet of contactor
out	refers to outlet of contactor
C	refers to cross-current extraction
G	refers to phase of flowrate G
L	refers to phase of flowrate L

Superscript

* denotes inversion

The above quantities may be expressed in any set of consistent units in which force and mass are not defined independently.

References

1. A. Kremser, Theoretical analysis of absorption process. *Nat. Pet. News* **22** (21), 42 (1930).
2. A. P. Colburn, The simplified calculation of diffusional processes. General consideration of two-film resistances, *Trans. Am. Inst. Chem. Engrs.* **35**, 211 (1939).
3. J. H. Perry (Editor), *Chemical Engineers' Handbook*, 3rd ed., p. 555, McGraw-Hill Book Company Incorporated, New York, 1950.
4. T. K. Sherwood and R. L. Pigford, *Absorption and Extraction*, p. 135, McGraw-Hill Book Company Incorporated, New York, 1952.
5. M. Souders and G. C. Brown, Fundamental design of absorbing and stripping columns for complex vapours, *Ind. Eng. Chem.* **24**, 519 (1932).
6. K. W. Warner, Extraction of uranyl nitrate in a disc column, *Chem. Eng. Sci.* **3**, 161 (1954).
7. L. D. Smoot and A. L. Babb, Mass transfer studies in a pulsed extraction column, *Ind. Eng. Chem. (Fundamentals)* **1** (2), 93 (1962).
8. S. Hartland, Design and Operation of Stagewise and Differential Counter-current Extractors. Ph.D. Thesis, University of Nottingham (1966).
9. S. Hartland, Calculation of numbers of stages and transfer units. *Trans. Inst. Chem. Engrs.* **44**, T116 (1966).
10. S. Hartland, The optimisation of forward and back extraction. Part I. Counter current flow. *Trans. Instn. Chem. Engrs. (London)* **45**, T82 (1967).
11. S. Hartland, The optimisation of forward and back extraction. Part II. Cross current flow with recirculation. *Trans. Instn. Chem. Engrs. (London)* **45**, T90 (1967).
12. S. Hartland, Optimum operating conditions in a forward and back extractor. *Chem. Eng. Sci.* **24**, 1075 (1969).
13. S. Hartland and V. Ang, Linear equilibrium data for ternary liquid systems and binary liquid–gas systems., *J. Chem. Eng. Data* **13**, 361 (1968).
14. L. Alders, *Liquid–Liquid Extraction*, 2nd ed., Elsevier Publishing Company, Amsterdam, 1959.

APPENDIX 1a
STAGEWISE CONCENTRATION PROFILES

Consider the stagewise contactor shown in Fig. A 1a.1 in which the number of theoretical stages n is measured from the y inlet. Using the fact that the x and y concentrations in each stage are related by the equilibrium expression 1(1), a solute balance round the typical nth stage leads to

$$y_{n+1} - (1 + J)\, y_n + J y_{n-1} = 0 \qquad \text{A1a(1)}$$

where $J = mG/L$. The solution to this finite difference equation is

$$y_n = \mathscr{A} J^n + \mathscr{B}$$

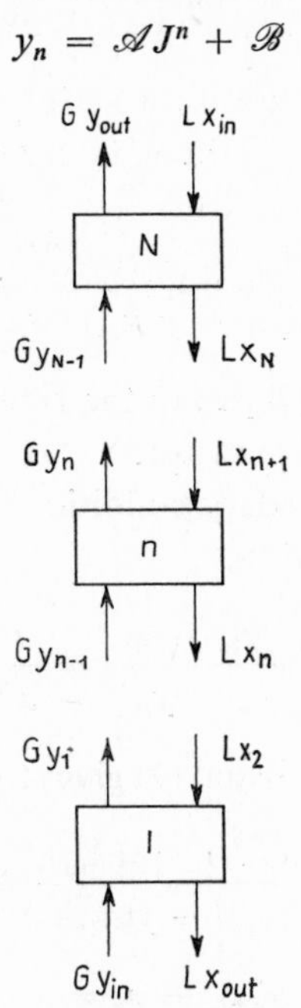

FIG. A1a.1. Flows and concentrations in a stagewise contactor.

and the two unknown constants $\mathscr{A}$ and $\mathscr{B}$ may be obtained from the boundary conditions

$$y_0 = y_{\text{in}} \quad \text{when} \quad n = 0 \qquad \text{A1a(2)}$$

$$x_1 = x_{\text{out}} \quad \text{when} \quad n = 1 \qquad \text{A1a(3)}$$

$$y_N = y_{\text{out}} \quad \text{when} \quad n = N \qquad \text{A1a(4)}$$

$$x_{N+1} = x_{\text{in}} \quad \text{when} \quad n = N + 1 \qquad \text{A1a(5)}$$

Because the stages are at equilibrium $y_n = y(x_n)$ and so the boundary conditions A1a(3) and (5) may be written

$$y_1 = y(x_{\text{out}}) \quad \text{when} \quad n = 1 \qquad \text{A1a(6)}$$

and

$$y_{n+1} = y(x_{\text{in}}) \quad \text{when} \quad n = N + 1 \qquad \text{A1a(7)}$$

Using boundary conditions A 1a(1) and A 1a(7), referring to the two inlet streams gives the dimensionless y concentration,

$$\phi_n = \frac{J(y_{\text{in}} - y_n)}{y_{\text{in}} - y(x_{\text{in}})} = \frac{J^n - 1}{J^N - 1/J} \qquad \text{A 1a(8)}$$

Dividing throughout by J and subtracting both sides of the equation from unity this may be rewritten in terms of the dimensionless x concentration,

$$\xi_{n+1} = \frac{y(x_{n+1}) - y(x_{\text{in}})}{y_{\text{in}} - y(x_{\text{in}})} = \frac{J^N - J^n}{J^N - 1/J} \qquad \text{A 1a(9)}$$

remembering that x_{n+1} and y_n are the concentrations of the streams passing each other in the contactor.

Alternatively using boundary conditions A 1a(2) and (4) gives the dimensionless y concentration,

$$\psi_n = \frac{y_{\text{in}} - y_n}{y_{\text{in}} - y_{\text{out}}} = \frac{J^n - 1}{J^N - 1} \qquad \text{A 1a(10)}$$

and using conditions A 1a(6) and (7) gives the dimensionless x concentration

$$\chi_{n+1} = \frac{y(x_{n+1}) - y(x_{\text{in}})}{y(x_{\text{out}}) - y(x_{\text{in}})} = \frac{J^N - J^n}{J^N - 1} \qquad \text{A 1a(11)}$$

so that

$$\chi_{n+1} + \psi_n = 1 \qquad \text{A 1a(12)}$$

Both χ and ψ always vary between 0 and 1.

When J is unity, equations A 1a(10) and (11) are indeterminate but writing $J = 1 + \delta$ and expanding shows they become

$$\psi_n = n/N \qquad \text{A 1a(13)}$$

and

$$\chi_{n+1} = 1 - n/N \qquad \text{A 1a(14)}$$

APPENDIX 1b
DIFFERENTIAL CONCENTRATION PROFILES

Consider the differential contactor of length H and cross-sectional area s shown in Fig. A 1b.1, in which distance h is measured from the y inlet. Equating the solute gained by the y stream, to that lost by the

x stream, both of which must be equal to the transfer from the x to y streams, leads to

$$\frac{\mathrm{d}y}{\mathrm{d}h} = \frac{1}{J}\frac{\mathrm{d}y(x)}{\mathrm{d}h} = \frac{y(x) - y}{H_G} \qquad \text{A1b(1)}$$

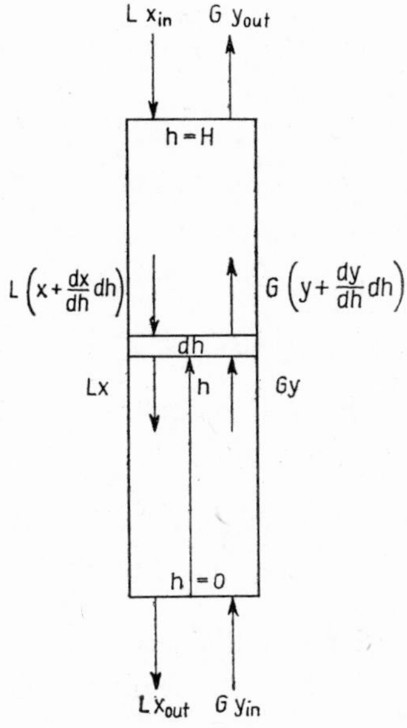

FIG. A1b.1. Flows and concentrations in a differential contactor.

In this expression

$$J = \frac{mG}{L} \quad \text{and} \quad H_G = \frac{G}{k_G as}$$

is the height of an overall transfer unit based on the G phase, and a the interfacial area per unit volume. Differentiation enables the concentrations y and $y(x)$ to be separated so that

$$\frac{\mathrm{d}^2y(x)}{\mathrm{d}h^2} = \frac{J-1}{H_G}\frac{\mathrm{d}y(x)}{\mathrm{d}h} \qquad \text{A1b(2)}$$

and

$$\frac{\mathrm{d}^2y}{\mathrm{d}h^2} = \frac{J-1}{H_G}\frac{\mathrm{d}y}{\mathrm{d}h}. \qquad \text{A1b(3)}$$

Integration of either of these expressions introduces two unknown constants which may be obtained from the boundary conditions

$$y = y_{\text{in}} \quad \text{at} \quad h = 0 \qquad \text{A1b(4)}$$

$$x = x_{\text{in}} \quad \text{at} \quad h = H \qquad \text{A1b(5)}$$

$$x = x_{\text{out}} \quad \text{at} \quad h = 0 \qquad \text{A1b(6)}$$

$$y = y_{\text{out}} \quad \text{at} \quad h = H \qquad \text{A1b(7)}$$

Using the inlet boundary conditions A1b(4) and (5) gives the dimensionless y concentration,

$$\phi = \frac{J(y_{\text{in}} - y)}{y_{\text{in}} - y(x_{\text{in}})} = \frac{e^{-N_G(1-J)h/H} - 1}{e^{-N_G(1-J)} - 1/J} \qquad \text{A1b(8)}$$

for it follows from equation A1b(1) that the x and y concentrations at any cross-section h are related by

$$\frac{Jy - y(x)}{J-1} = \frac{Jy_{\text{in}}\, e^{-N_G(1-J)} - y(x_{\text{in}})}{J\, e^{-N_G(1-J)} - 1} \qquad \text{A1b(9)}$$

Using this latter equation to eliminate y from equation A1b(5) remembering that $N_L = JN_G$ leads to the dimensionless x concentration,

$$\xi = \frac{y(x) - y(x_{\text{in}})}{y_{\text{in}} - y(x_{\text{in}})} = \frac{e^{N_L(1-1/J)} - e^{N_L(1-1/J)h/H}}{e^{N_L(1-1/J)} - 1/J} \qquad \text{A1b(10)}$$

Alternatively using the boundary conditions A1b(4) and (7) gives the dimensionless y concentration

$$\psi = \frac{y_{\text{in}} - y}{y_{\text{in}} - y_{\text{out}}} = \frac{e^{-N_G(1-J)h/H} - 1}{e^{-N_G(1-J)} - 1} \qquad \text{A1b(11)}$$

and using conditions A1b(5) and (6) gives the dimensionless x concentration

$$\chi = \frac{y(x) - y(x_{\text{in}})}{y(x_{\text{out}}) - y(x_{\text{in}})} = \frac{e^{N_L(1-1/J)} - e^{N_L(1-1/J)h/H}}{e^{N_L(1-1/J)} - 1} \qquad \text{A1b(12)}$$

so that

$$\psi + \chi = 1 \qquad \text{A1b(13)}$$

When J is unity the equations are indeterminate but writing $J = 1 + \delta$ and expanding they become

$$\psi = h/H \qquad \text{A1b(14)}$$

and

$$\chi = 1 - h/H \qquad \text{A1b(15)}$$

Equations A1b(8), (10), (11) and (12) express the concentrations y and x in terms of distance h.

APPENDIX 1c
CROSS-CURRENT EXTRACTION

The extraction of a solvent G with a cross-current solvent L is illustrated in Fig. A1c.1. When equilibrium is reached in each stage a

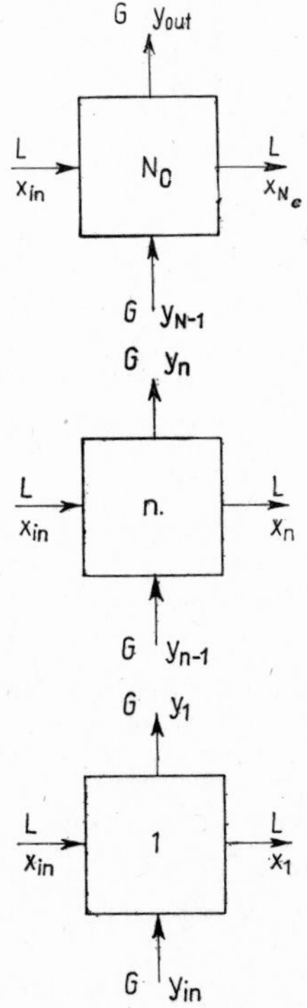

FIG. A1c.1. Flows and concentrations for cross-current flow.

solute balance round a typical stage n leads to

$$(1 + J)\, y_n - J y_{n-1} = y(x_{\text{in}}) \qquad \text{A 1c(1)}$$

where

$$J = mG/L$$

and

$$y(x_{\text{in}}) = m x_{\text{in}} + c$$

is the y concentration in equilibrium with the cross-current solvent feed. This finite difference equation has solution

$$y_n = \mathscr{A}\lambda^n + \mathscr{B} \qquad \text{A 1c(2)}$$

where

$$\lambda = \frac{J}{1 + J}$$

and $\mathscr{A}$ and $\mathscr{B}$ are constants which may be determined from the boundary conditions. These are

$$n = 0 \quad \text{when} \quad y_n = y_{\text{in}} \qquad \text{A 1c(3)}$$

$$n = N_C \quad \text{when} \quad y_n = y_{\text{out}} \qquad \text{A 1c(4)}$$

and

$$n = 1 \quad \text{when} \quad y_n = y_{\text{in}} + y(x_{\text{in}}) \qquad \text{A 1c(5)}$$

which may be obtained from a balance around the first stage. Using conditions A 1c(3) and (4) leads to

$$\frac{y_{\text{in}} - y_n}{y_{\text{in}} - y_{\text{out}}} = \frac{1 - \lambda^n}{1 - \lambda^{N_C}} \qquad \text{A 1c(6)}$$

which is the concentration profile of the y phase. Alternatively using conditions A 1c(3) and (5) leads to

$$\frac{y_n - y(x_{\text{in}})}{y_{\text{in}} - y(x_{\text{in}})} = \lambda^n \qquad \text{A 1c(7)}$$

which involves the inlet concentrations which are more likely to be known.

Introducing the condition A 1c(4) gives the overall separation factor

$$\frac{y_{\text{out}} - y(x_{\text{in}})}{y_{\text{in}} - y(x_{\text{in}})} = \lambda^{N_C} \qquad \text{A 1c(8)}$$

This may be re-expressed

$$N_C \ln 1/\lambda = \ln Q_C \qquad \text{A 1c(9)}$$

where

$$Q_C = \frac{y_{\text{in}} - y(x_{\text{in}})}{y_{\text{out}} - y(x_{\text{in}})}$$

and the subscript C refers to cross-current extraction. Cross-current extraction may thus be represented by Fig. A1c.2 which relates N_C and Q_C at different values of the extraction factor J. This shows that as J becomes infinite and λ unity, an infinite number of stages are required for a given extraction. However, as J and hence λ approach zero so also does N_C.

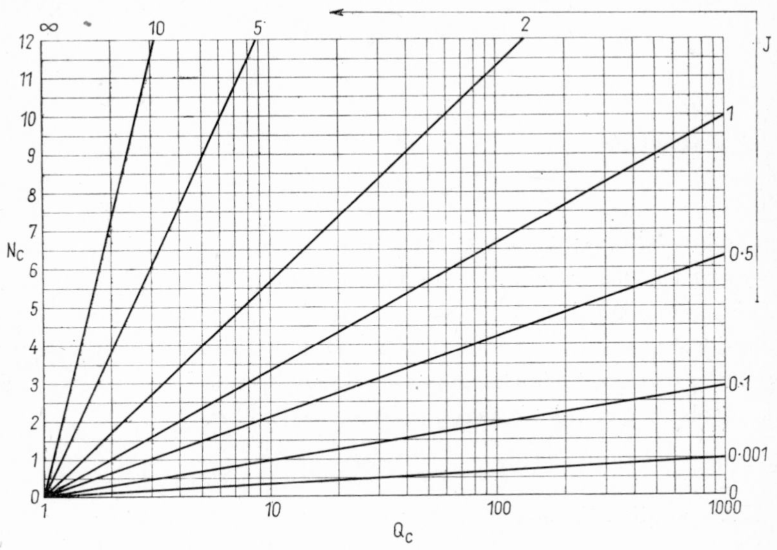

FIG. A1c.2. Variation of number of cross-current stages N_C with separation Q_C for different values of extraction factor J.

When there is no solute in the cross-current feed and the equilibrium line passes through the origin (so x_{in} and the intercept c are both zero) equation A1c(8) reduces to

$$y_{out}/y_{in} = \lambda^{N_C} \qquad \text{A1c(10)}$$

This form of expression involving the fraction of unextracted solute was derived by Alders.[14]

CHAPTER 2

Formulae Relating *N* and *J* to the Different Separation Factors *Q*

2.1. Introduction

Many separation factors are possible in addition to those defined by equations 1(3) and 1(6), and many relations between *N*, *J* and *Q* have appeared in the literature (references 1–9), some of them conflicting. This chapter considers all possible separation factors and derives their relationship to the extraction factor and number of stages or transfer units. It attempts to find the most convenient formula for general use and to resolve the conflict between published formulae.

2.2. Separation Factors

There are fourteen basic different separation factors as the following analysis shows. Each separation factor is the ratio of two concentration differences. There are four end concentrations x_{in}, x_{out}, y_{in}, y_{out} as indicated in Fig. 1.1, and hence each concentration difference is the result of choosing two concentrations from four.

Now

$$^2C_4 = \frac{4!}{2!\,2!} = 6$$

and hence there are six different numerator combinations and the same six denominator combinations. The number of ways of choosing a numerator and a denominator from the six possibilities is

$$^2C_6 = \frac{6!}{4!\,2!} = 15.$$

However, this includes

$$\frac{y_{in} - y_{out}}{y(x_{out}) - y(x_{in})}$$

which from an overall balance is equal to J^* and hence the possible number of different separation factors is only fourteen. The method of analysis automatically excludes negative and reciprocal variations. If these are included there are $2 \times 2 \times 14 = 56$ separation factors.

The six possible combinations of the end concentrations are listed in Table 2.1 together with the fourteen possible basic separation factors. These may be interrelated by simply adding and subtracting concentrations and by using the overall balance. Thus all the separation factors may be expressed in terms of the separation factor Q of equation 1(3) (this is $\Delta y_2/\Delta y_4$ of Table 2.1 which we will write $Q_{2/4}$) and these relationships substituted into equations 1(4) and 1(5) to obtain new expressions for N. (The interrelation of Q and Q^* discussed in Chapter 1 is one example.) Separation factors containing Δy_6 are very simply related to those containing Δy_1, for these two concentration differences are related by the overall balance 1(9) as pointed out above.

TABLE 2.1. SEPARATION FACTORS

Concentration difference	Separation factors
$\Delta y_1 = y_{in} - y_{out}$	$\frac{\Delta y_1}{\Delta y_2}, \frac{\Delta y_1}{\Delta y_3}, \frac{\Delta y_1}{\Delta y_4}, \frac{\Delta y_1}{\Delta y_5}, \left[\frac{\Delta y_1}{\Delta y_6} = J^*\right]$
$\Delta y_2 = y_{in} - y(x_{in})$	$\frac{\Delta y_2}{\Delta y_3}, \frac{\Delta y_2}{\Delta y_4}, \frac{\Delta y_2}{\Delta y_5}, \frac{\Delta y_2}{\Delta y_6}$
$\Delta y_3 = y_{in} - y(x_{out})$	$\frac{\Delta y_3}{\Delta y_4}, \frac{\Delta y_3}{\Delta y_5}, \frac{\Delta y_3}{\Delta y_6}$
$\Delta y_4 = y_{out} - y(x_{in})$	$\frac{\Delta y_4}{\Delta y_5}, \frac{\Delta y_4}{\Delta y_6}$
$\Delta y_5 = y(x_{out}) - y_{out}$	$\frac{\Delta y_5}{\Delta y_6}$
$\Delta y_6 = y(x_{out}) - y(x_{in})$	

2.3. Inversion

If Fig. 1.1 is physically inverted it may be seen that L, G, x_{in}, x_{out}, y_{in} and y_{out} may be replaced respectively by G, L, y_{in}, y_{out}, x_{in} and x_{out} for the x and y streams are interchanged. The equilibrium relationship 1(1) may be written

$$x = y/m - c/m$$

and so it follows that m and c may be replaced by $1/m$ and $-c/m$. Thus $J = mG/L$ may be replaced by $L/(Gm) = 1/J = J^*$ and, for example, the equilibrium concentration $y(x_{in})$ by $x(y_{in})$. Denoting inversion by an asterisk (*) the separation factor $Q_{2/4}$ defined by equation 1(3) may

TABLE 2.2. INVERSE QUANTITIES IN COUNTER-CURRENT EXTRACTION

Quantity	Inverse
L	G
G	L
x	y
x_n	y_n
y	x
y_n	x_n
x_{in}	y_{in}
y_{in}	x_{in}
x_{out}	y_{out}
y_{out}	x_{out}
m	$1/m$
c	$-c/m$
$J = \dfrac{mG}{L}$	$J^* = \dfrac{L}{mG}$
N_S	N_S
k_G	k_L
k_L	k_G
$N_G = \dfrac{k_G A}{G}$	$N_L = \dfrac{k_L A}{L}$
$N_L = \dfrac{k_L A}{L}$	$N_G = \dfrac{k_G A}{G}$
n	$N - n + 1$
h	$H - h$

The separation factors Q and their inverses Q^* are defined in Table 2.3.

thus be replaced by

$$Q^*_{2/4} = \frac{x_{\text{in}} - x(y_{\text{in}})}{x_{\text{out}} - x(y_{\text{in}})} = Q_{2/3}$$

which is identical with equation 1(6).

The number of stages N_S is unchanged on inversion but N_G is replaced by N_L for the overall mass transfer coefficient k_G becomes k_L.

Inversion of Figs. A1a.1 and A1b.1 shows that the number of stages, n is replaced by $(N - n + 1)$ and distance, h by $(H - h)$. Use of these relationships together with those discussed above enables the stagewise concentration profiles 1(15) and 1(16) and the differential concentration profiles 1(21) and 1(22) to be interconverted. This means that χ and ψ may be interchanged in Figs. 1.5 and 1.6 if J is replaced by $1/J$, n by $N - n + 1$, h by $H - h$ and N_G by N_L. The figures may thus also be used when $J = 1{\cdot}43$ and in general such figures need only cover the region J less than one. The stagewise profiles 1(12) and 1(13) and the differential profiles 1(18) and 1(19) may be similarly inverted if ϕ is replaced by ξ/J and ξ by ϕ/J. A list of inverse quantities is given in Table 2.2.

2.4. Formulae Relating N, J and Q

All the separation factors may be inverted in the manner discussed above to give a separation factor already included in Table 2.1. Separation factors and their inverses are paired in Table 2.3. It is interesting to note that the inverse of $Q_{2/5}$ is $Q_{2/5}$, that is unchanged and the inverse of $Q_{3/4}$ is $Q_{4/3}$, that is the reciprocal. If the separation factors are expressed in terms of $Q_{2/4}$ and substituted into equations 1(4) and 1(5) the equations for N_S and N_G also given in Table 2.3 are obtained. Identical equations for N_S and N_L, but now involving J^* instead of J, are obtained if the separation factors Q are replaced by their inverses Q^*. We have already seen in Chapter 1 how equations 1(4) and 1(5) relating N, J and $Q_{2/4}$ are identical in form with equations 1(10) and 1(11) relating N, J^* and $Q_{2/3}$, and another example is given in Appendix 2a. There are thus only eight basic forms of the equations relating N, J and Q even though there are fourteen basic separation factors.

Values of N_S, N_G and N_L when $J = J^* = 1$ were obtained by writing $J = 1 + \delta$, expanding and letting δ approach zero, and are given in Table 2.3. Alternatively all the formulae in Table 2.3 may be derived from basic principles as shown for the stagewise case in Appendix 2b. Examples of the use of the formulae are given in Appendix 2c.

TABLE 2.3. FORMULAE RELATING N, J AND Q OR N, J^* AND Q^*

N_S and N_L may be obtained by replacing J and Q by J^* and Q^* in the formulae for N_S and N_G given below.

Q	Q^*	Range for $N_S > 1$	$\left(\frac{1}{J}\right)^{N_S}$ or $\exp[N_G(1-J)]$	$N_S = N_G = N_L$ at $J = J^* = 1$	Formula number
$\frac{\Delta y_1}{\Delta y_2}$	$\frac{\Delta y_6}{\Delta y_2}$	0 to 1	$\frac{1-JQ}{1-Q}$	$\frac{Q}{1-Q}$	I
$\frac{\Delta y_1}{\Delta y_3}$	$\frac{\Delta y_6}{\Delta y_4}$	0 to ∞	$\frac{1}{1+Q(J-1)}$	Q	II
$\frac{\Delta y_1}{\Delta y_4}$	$\frac{\Delta y_6}{\Delta y_3}$	0 to ∞	$1+Q(1-J)$	Q	III
$\frac{\Delta y_1}{\Delta y_5}$	$\frac{\Delta y_6}{\Delta y_5}$	0 to ∞	$\frac{1-Q}{1-JQ}$	$\frac{Q}{Q-1}$	IV
$\frac{\Delta y_2}{\Delta y_4}$	$\frac{\Delta y_2}{\Delta y_3}$	1 to ∞	$(1-J)Q+J$	$Q-1$	V
$\frac{\Delta y_2}{\Delta y_5}$	$\frac{\Delta y_2}{\Delta y_5}$	1 to ∞	$\frac{J-Q}{1-JQ}$	$\frac{Q+1}{Q-1}$	VI
$\frac{\Delta y_3}{\Delta y_4}$	$\frac{\Delta y_4}{\Delta y_3}$	1 to ∞	Q	Indeterminate	VII
$\frac{\Delta y_5}{\Delta y_3}$	$\frac{\Delta y_5}{\Delta y_4}$	0 to ∞	$\frac{1}{J+Q(J-1)}$	$Q+1$	VIII

The range of the separation factors is of interest and is given in Table 2.3. This may be obtained by considering Fig. 1.4 which gives the relative positions of the end concentrations for absorption and stripping for the two cases y_{out} less than $y(x_{out})$ which is usual in absorption and y_{out} greater than $y(x_{out})$ which is usual in stripping. If y_{out} is greater than $y(x_{out})$ in absorption or y_{out} is less than $y(x_{out})$ in stripping it corresponds to N_S being less than one. From Fig. 1.4 it can be seen that the range of any separation factor is the same as that of its inverse. If N_S is greater than one, all the separation factors and their inverses are positive.

2.5. Comparison with Other Formulae

It is interesting to compare the formulae in Table 2.3 with those derived previously in the literature. The following comparisons are for the stagewise case.

Souders and Brown[3] derived:

$$\frac{y_{\text{in}} - y_{\text{out}}}{y_{\text{in}} - x_{\text{in}}} = \frac{(1/J)^{N+1} - 1/J}{(1/J)^{N+1} - 1} \qquad 2(1)$$

which is formula I of Table 2.3, explicit in $Q_{1/2}$.

Underwood[4] derived:

$$x_{\text{out}} = \frac{x_{\text{in}}(J-1) + y_{\text{in}}(J^{N+1} - J)}{J^{N+1} - 1} \qquad 2(2)$$

which may be reduced to formula III of Table 2.3.

Hunter and Nash[5] derived:

$$x(y_{\text{out}}) = \frac{(J^N - 1)\, x_{\text{in}} + x(y_{\text{in}})(J-1) J^N}{(J-1) - J(J^N - 1)} \qquad 2(3)$$

which does not agree with formula I of Table 2.3.

Introducing the finite difference technique Tiller and Tour[6] derived:

$$y_n(1 - J) = (y_{\text{in}} - y(x_{\text{out}}))\, J^n + (y(x_{\text{in}}) - J y_{\text{out}}) \qquad 2(4)$$

when $n = N$, $y_n = y_{\text{out}}$ and the equation becomes identical with formula VII of Table 2.3.

The fourth edition of the *Chemical Engineers' Handbook*[7] quotes

$$-Q_{4/5} = \frac{x_{\text{in}} - x(y_{\text{out}})}{x_{\text{out}} - x(y_{\text{out}})} = \frac{J - 1}{J^{N+1} - 1} \qquad 2(5)$$

and

$$N_S = \frac{\ln [(1 - J^*)\, Q_{5/4} + J^*]}{\ln 1/J^*} \qquad 2(6)$$

which are not self-consistent. Neither formula agrees with Table 2.3.

The former gives negative values of N_S as may be seen by comparison with formula VIII.

One of the formulae suggested by Sherwood and Pigford[8] and the third edition of the *Chemical Engineers' Handbook*[9] for the stagewise and differential cases, is:

$$N_S \ln 1/J^* = N_G(1 - J^*) = \ln[(1 - J^*)Q^* + J^*] \qquad 2(7)$$

where

$$Q^* = \frac{x_{out} - x(y_{out})}{x_{in} - x(y_{out})} \qquad 2(8)$$

and this also leads to negative values of N.

Examples of the use of these latter formulae are given in Appendix 2c.

The differential formula VII involving $Q_{3/4}$ may be converted to that for log mean driving force based on the y phase, Δy_{lm}. Substituting for J from the overall balance 1(9) gives:

$$N_G = \frac{y_{in} - y_{out}}{\Delta y_{lm}} \qquad 2(9)$$

This equation is well known and the analogous equation is widely used in heat transfer.

2.6. Analogy with Heat Transfer

Suppose in Fig. 1.1 we replace the concentrations x_{in}, x_{out}, y_{in}, y_{out} by the temperatures $\theta_{L,in}$, $\theta_{L,out}$, $\theta_{G,in}$, $\theta_{G,out}$, respectively, so that the contactor becomes a heat exchanger. Using an overall balance and the definition of log mean temperature difference gives:

$$W_G \ln \frac{\theta_{G,in} - \theta_{L,out}}{\theta_{G,out} - \theta_{L,in}} = UA(1 - W_G/W_L) \qquad 2(10)$$

where when L and G are volume flowrates, $W_G = Gd_GC_G$ and $W_L = Ld_LC_L$, d is the density, C the specific heat, U the overall heat transfer coefficient and A the total area of contact between phases.

Thus writing $J = W_G/W_L$, $Q_{3/4} = \dfrac{\theta_{G,in} - \theta_{L,out}}{\theta_{G,out} - \theta_{L,in}}$ and $N_G = \dfrac{UA}{W_G}$ equation 2(10) becomes:

$$\ln Q_{3/4} = (1 - J)N_G$$

which is the differential formula VII of Table 2.3.

It has been shown earlier that any formula in Table 2.3 can be obtained from any other and so all the formulae may be used for heat transfer if the identities listed in Table 2.4 are used. These have been pointed out by Miyauchi and Vermeulen[(10)] for the more complicated case involving longitudinal dispersion. Kays and London[(11)] use the concept of heat transfer units and derive formulae similar to those in Table 2.3. Examples of the use of formula VII in heat transfer are given in Appendix 2d.

TABLE 2.4. ANALOGOUS QUANTITIES IN MASS AND HEAT TRANSFER

Mass transfer	Heat transfer
x	θ_L
y	θ_G
m	1
c	0
L	W_L
G	W_G
k_G	U
k_L	U
$J = \dfrac{mG}{L}$	$\dfrac{W_G}{W_L}$
$J^* = \dfrac{L}{mG}$	$\dfrac{W_L}{W_G}$
$N_G = \dfrac{k_G A}{G}$	$\dfrac{UA}{W_G}$
$N_L = \dfrac{k_L A}{L}$	$\dfrac{UA}{W_L}$

2.7. Discussion

The values of N_S and N_T for each of the formulae given in Table 2.3 have been calculated on an Atlas digital computer for over fifty values of J between 10^{-3} and 10 and over fifty values of Q between 10^{-6} to 10^6. Most of the plots relating N, J and the different separation factors suffer from the same faults as Figs. 1.2 and 1.3; the curves become asymptotic and are unevenly graded with respect to the constant parameter. This unfortunately applies to $Q_{2/5}$ which is unchanged on inversion.

However, the graphs involving $Q_{3/4}$ are very easy to use because of the simple relationship between $Q_{3/4}$, J and N_S or N_T. Plotted as in Figs. 2.1 and 2.2 they are both straight lines with for the most part an even gradation between the values of the constant parameter. Only the

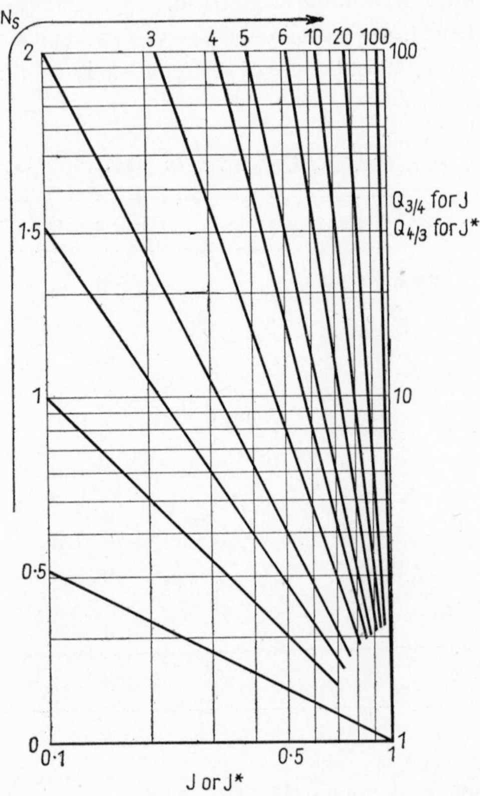

FIG. 2.1. Variation of $Q_{3/4}$ (or $Q_{4/3}$) with J (or J^*) for different values of N_S.

region J less than one is shown; when J is greater than one, its inverse $1/J$ and the inverse separation factor $Q_{4/3}$ should be used. Plots of N versus ln Q at constant values of J are also linear.

The formulae involving $Q_{3/4}$ does have the disadvantage that N is indeterminate when $J = 1$. This is because $Q_{3/4}$ is always one when $J = 1$, and is not affected by the length of the operating line. However, this disadvantage also applies to log mean concentration and temperature differences which are widely used. When $J = 1$ one of the other formulae must be used. Again some of the separation factors are more convenient to use than others in certain circumstances. For example,

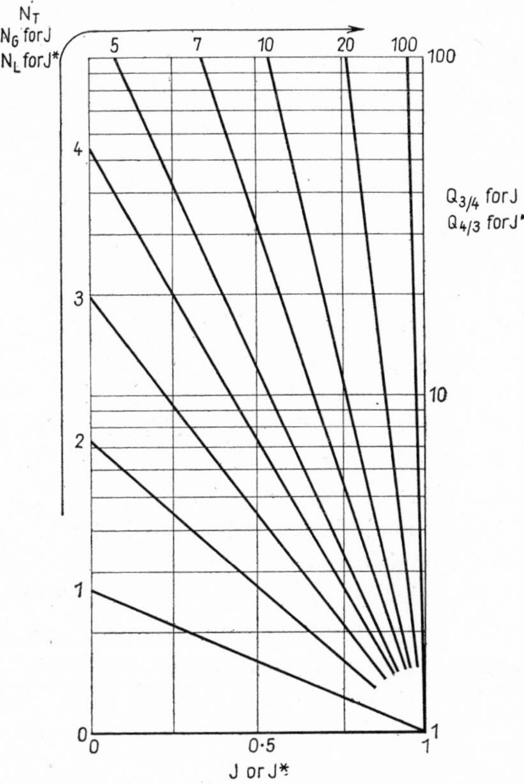

FIG. 2.2. Variation of $Q_{3/4}$ (or $Q_{4/3}$) with J (or J^*) for different values of N_T.

$Q_{2/4}$ defined by equation 1(3) is equal to y_{in}/y_{out} for absorption with $x_{in} = 0$ and $Q_{2/3}$, defined by equation 1(6), reduces to x_{in}/x_{out} for stripping with $y_{in} = c = 0$. Appendices 1a and 1b discuss how concentration profiles are most conveniently expressed. In addition most separation factors contain only three boundary concentrations whereas $Q_{3/4}$ contains all four which may necessitate the use of the overall balance 1(9).

However, on balance it would seem that formulae VII of Table 2.3

$$N_S \ln 1/J = N_G (1 - J) = \ln Q_{3/4} \qquad 2(11)$$

$$N_S \ln 1/J^* = N_L (1 - J^*) = \ln Q_{4/3} \qquad 2(12)$$

are the most convenient formulae for general use and it is recommended that they be used for the calculation of numbers of stages and transfer units. Some numerical examples based on these formulae are given below in Examples 2.1 and 2.2.

Example 2.1

This is taken from the third edition of the *Chemical Engineers' Handbook*[(9)] where $Q_{2/4}$ is used to perform the same calculation.

Calculate the number of transfer units required in a packed column used to absorb acetone from air by water. The gas mixture fed to the column contains 7 mole per cent acetone and 93 mole per cent air nearly saturated with water vapour at 25 °C and 1 atm. The water fed to the top of the column is free of acetone and its temperature is 25 °C. The exit gas is to contain only 0·01 mole per cent acetone and sufficient water is to be supplied to the column so that $J = mG/L = 0{\cdot}5$ at the dilute end where most of the transfer units are used. At 25 °C and 1 atm total pressure, the vapour pressure of acetone over its dilute water solution corresponds to $m = 1{\cdot}75$.

SOLUTION

Assuming isothermal operation, the equilibrium line is straight throughout the column and the operating line is substantially straight. If air is taken as the G phase and water the L phase then $y_{\text{in}} = 7$, $y_{\text{out}} = 0{\cdot}01$ and $x_{\text{in}} = 0$. $L/G = m/J = 1{\cdot}75/0{\cdot}5 = 3{\cdot}5$ and so from an overall balance $x_{\text{out}} = (7 - 0{\cdot}01)/3{\cdot}5 = 2$.

Thus $Q_{3/4} = \dfrac{y_{\text{in}} - y(x_{\text{out}})}{y_{\text{out}} - y(x_{\text{in}})} = \dfrac{7 - 2 \times 1{\cdot}75}{0{\cdot}01 - 0} = 350$ and so $N_G = \ln Q_{3/4}/(1 - J) = \ln 350/0{\cdot}5 = 11{\cdot}7$ which could also be obtained from Fig. 2.2.

The number of theoretical stages required to give the same separation with the same operating conditions is

$$N_S = \ln Q_{3/4}/\ln 1/J = \ln 350/\ln 2 = 8 \cdot 45$$

which could also be obtained from Fig. 2.1.

Example 2.2

This is taken from Treybal[12] where $Q_{2/4}$ is used to perform the same calculation.

Leibson and Beckmann[13] extracted diethylamine from water ($20 \cdot 44$ ft^3/hr ft^2) with toluene ($3 \cdot 05$ ft^3/hr ft^2) at $30 \cdot 8$ °C, toluene dispersed, in a 6-in. inside diameter tower packed to a depth of 4 ft with $\frac{1}{2}$ in. Raschig rings. The observed concentrations of diethylamine, in lb moles/ft^3 were water: in = 0·01574, out = 0·01450; toluene: in = 0 out = 0·00860. At this temperature the distribution coefficient (concentration in water/concentration in toluene) is equal to 1·156, and water and toluene are substantially insoluble at these diethylamine concentrations. Compute the effluent concentrations if 8 ft of packing had been used.

SOLUTION

If there are N transfer units in 4 ft of packing there will be $2N$ in 8 ft of packing and J will be the same in both cases as the operating conditions are unchanged. Dividing the two equations for N it thus follows that $Q^2_{4\,\text{ft}} = Q_{8\,\text{ft}}$ where Q is of the form $Q_{3/4}$.

Now

$$Q_{4\,\text{ft}} = \frac{0 - 0 \cdot 0145/1 \cdot 156}{0 \cdot 00860 - 0 \cdot 01574/1 \cdot 156} = 2 \cdot 5$$

and so

$$Q_{8\,\text{ft}} = \frac{0 - x_{\text{out}}/1 \cdot 156}{y_{\text{out}} - 0 \cdot 0136} = 6 \cdot 25\,.$$

Thus $6 \cdot 25 y_{\text{out}} - 0 \cdot 085 = -x_{\text{out}}/1 \cdot 156$ and from an overall balance $0 \cdot 01574 - x_{\text{out}} = 3 \cdot 05 y_{\text{out}}/20 \cdot 4$, which gives $y_{\text{out}} = 0 \cdot 0117$ and $x_{\text{out}} = 0 \cdot 0140$.

For a stagewise process giving the same separation and with the same operating conditions, doubling the number of stages would produce the same result.

Notation for Chapter 2

A	total area of contact between two phases
c	intercept of equilibrium line on y-axis
C	specific heat
d	density
G	flowrate of phase in which solute concentration is y
J	extraction factor ($J = mG/L$)
J^*	inverse extraction factor ($J^* = L/mG$)
L	flowrate of phase in which solute concentration is x
m	slope of equilibrium line (equation $y = mx + c$)
N	number of theoretical stages or transfer units
N_G	number of overall transfer units based on phase of flowrate G
N_L	number of overall transfer units based on phase of flowrate L
N_S	number of theoretical stages
N_T	number of transfer units
Q	separation factor (defined by Table 2.1)
Q^*	inverse separation factor (defined by Table 2.3)
U	overall heat transfer coefficient
W	heat capacity of stream per unit time
x	solute concentration in phase of flowrate L
$x(y)$	x concentration in equilibrium with concentration y ($x(y) = (y - c)/m$)
y	solute concentration in phase of flowrate G
$y(x)$	y concentration in equilibrium with concentration x ($y(x) = mx + c$)

Greek symbols

θ	temperature
Δy	concentration difference
Δy_{lm}	log mean concentration difference based on the y phase
$\Delta \theta_{\text{lm}}$	log mean temperature difference

Subscripts

in	refers to inlet of contactor
out	refers to outlet of contactor

G refers to phase of flowrate G
L refers to phase of flowrate L

Superscript
* denotes inversion

The above quantities may be expressed in any set of consistent units in which force and mass are not defined independently.

References

1. A. Kremser, Theoretical analysis of absorption process. *Nat. Pet. News* **22,** (21) 42 (1930).
2. A. P. Colburn, The simplified calculation of diffusional processes. General consideration of two-film resistances, *Trans. Am. Inst. Chem. Engrs.* **35,** 211 (1939).
3. M. Souders and G. C. Brown, Fundamental design of absorbing and stripping columns for complex vapours, *Ind. Eng. Chem.* **24,** 519 (1932).
4. A. J. V. Underwood, Graphical calculation for extraction problems, *Ind. Eng. Chem.* **10,** 128 (1934).
5. T. G. Hunter and A. W. Nash, The application of physico-chemical principles to the design of liquid–liquid contact equipment. Part I. General theory, *J. Soc. Chem. Ind.* **51,** 285T (1932).
6. F. M. Tiller and R. S. Tour, Stagewise operations—Application of the calculus of finite differences to chemical engineering, *Trans. Am. Inst. of Chem. Engrs.* **40,** 317 (1944).
7. J. H. Perry, C. H. Chilton and S. D. Kirkpatrick (Editors), *Chemical Engineers' Handbook*, 4th ed., pp. 14–60, McGraw-Hill Book Company Incorporated, New York, 1963.
8. T. K. Sherwood and R. L. Pigford, *Absorption and Extraction*, p. 135, McGraw-Hill Book Company Incorporated, New York, 1952.
9. J. H. Perry (Editor), *Chemical Engineers' Handbook*, 3rd ed., p. 555, McGraw-Hill Book Company Incorporated, New York, 1950.
10. T. Miyauchi and T. Vermeulen, Longitudinal dispersion in two-phase continuous flow operations, *Ind. Eng. Chem. (Fundamentals)* **2** (2), 113 (1963).
11. W. M. Kays and A. L. London, *Compact Heat Exchangers*, 2nd ed., McGraw-Hill Book Company Incorporated, New York, 1964.
12. R. E. Treybal, *Mass Transfer Operations*, p. 432, McGraw-Hill Book Company Incorporated, New York, 1955.
13. I. Leibson and R. B. Beckmann, The effect of packing size and column diameter on mass transfer in liquid–liquid extraction, *Chem. Eng. Progr.* **49,** 405 (1953).

APPENDIX 2a
INVERSION OF FORMULAE RELATING N, J AND Q

Consider the separation factor $Q_{2/5}$ which is defined by

$$Q_{2/5} = \frac{y_{\text{in}} - y(x_{\text{in}})}{y(x_{\text{out}}) - y_{\text{out}}} \qquad \text{A2a(1)}$$

the modulus of which is always greater than one. The relationship between $Q_{2/5}$ and $Q_{2/4}$ defined by equation 1(3) is

$$Q_{2/4} = -\frac{(1 + J)\,Q_{2/5}}{1 - JQ_{2/5}} \qquad \text{A2a(2)}$$

and when substituted into equations 1(4) and 1(5) this gives

$$N_S \ln 1/J = N_G\,(1 - J) = \ln\frac{J - Q_{2/5}}{1 - JQ_{2/5}} \qquad \text{A2a(3)}$$

Inversion of equation A2a(1) followed by interchange of the equilibrium concentrations gives

$$Q^*_{2/5} = \frac{x_{\text{in}} - x(y_{\text{in}})}{x(y_{\text{out}}) - x_{\text{out}}} = \frac{y_{\text{in}} - y(x_{\text{in}})}{y(x_{\text{out}}) - y_{\text{out}}} = Q_{2/5}$$

which shows that the separation factor $Q_{2/5}$ is unchanged on inversion. Replacing N_S, N_G, J and $Q_{2/5}$ by N_S, N_L, J^* and $Q_{2/5}$ respectively in equation A2a(3) it follows directly that

$$N_S \ln 1/J^* = N_L\,(1 - J^*) = \ln\frac{J^* - Q_{2/5}}{1 - J^*Q_{2/5}}$$

APPENDIX 2b
DERIVATION OF FORMULAE RELATING N, J AND Q FROM BASIC PRINCIPLES

Consider the stagewise case, for example, as shown in Fig. A2b.1. The direction in which the stages are numbered may be used to introduce the inverse relationships. Deriving the equation in terms of x rather than y merely replaces each concentration by its equilibrium value.

Numbering from y inlet *Numbering from x inlet*

Using the fact that the stages are at equilibrium a balance round the typical nth stage gives the finite difference equations

$$y_{n+1} - (1 + J)\,y_n + Jy_{n-1} = 0 : y_{n+1} - (1 + J^*)\,y_n + J^*y_{n-1} = 0$$

which have solution

$$y_n = \mathscr{A}_1 J^n + \mathscr{A}_2 \qquad : y_n = \mathscr{B}_1 J^{*n} + \mathscr{B}_2$$

where $\mathscr{A}_1, \mathscr{A}_2, \mathscr{B}_1$ and $\mathscr{B}_2$ are unknown constants which may be determined using the boundary conditions,

$$y_{in} = y_0 = \mathscr{A}_1 + \mathscr{A}_2 \qquad : y(x_{in}) = y_0 = \mathscr{B}_1 + \mathscr{B}_2$$

$$y_{out} = y_N = \mathscr{A}_1 J^N + \mathscr{A}_2 \qquad : y(x_{out}) = y_N = \mathscr{B}_1 J^{*N} + \mathscr{B}_2$$

$$y(x_{in}) = y_{N+1} = \mathscr{A}_1 J^{N+1} + \mathscr{A}_2 : y_{in} = y_{N+1} = \mathscr{B}_1 J^{*N+1} + \mathscr{B}_2$$

$$y(x_{out}) = y_1 = \mathscr{A}_1 J + \mathscr{A}_2 \qquad : y_{out} = y_1 = \mathscr{B}_1 J^* + \mathscr{B}_2$$

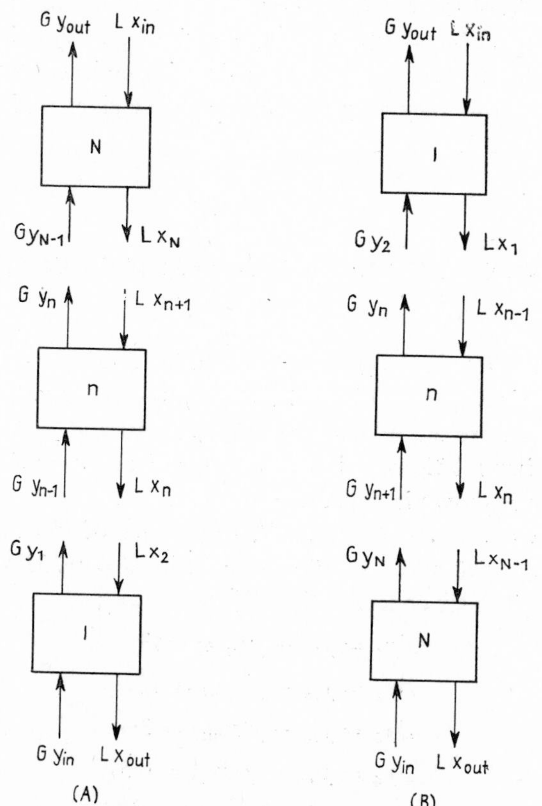

FIG. A2b.1. Flows and concentrations in a stagewise contactor. Numbering stages (A) from y inlet and (B) from x inlet.

Subtracting to eliminate the second constant, $\mathscr{A}_2$ or $\mathscr{B}_2$

$$\Delta y_1 = y_{in} - y_{out} = \mathscr{A}_1(1 - J^N) \ : \ -\Delta y_6 = x_{in} - x_{out} = \mathscr{B}_1(1 - J^{*N})$$
$$\Delta y_2 = y_{in} - x_{in} = \mathscr{A}_1(1 - J^{N+1}) \ : \ -\Delta y_2 = x_{in} - y_{in} = \mathscr{B}_1(1 - J^{*N+1})$$
$$\Delta y_3 = y_{in} - x_{out} = \mathscr{A}_1(1 - J) \ : \ -\Delta y_4 = x_{in} - y_{out} = \mathscr{B}_1(1 - J^*)$$
$$\Delta y_4 = y_{out} - x_{in} = \mathscr{A}_1(J^N - J^{N+1}) : -\Delta y_3 = x_{out} - y_{in} = \mathscr{B}_1(J^{*N} - J^{*N+1})$$
$$\Delta y_5 = x_{out} - y_{out} = \mathscr{A}_1(J - J^N) \ : \ -\Delta y_5 = y_{out} - x_{out} = \mathscr{B}_1(J^* - J^{*N})$$
$$\Delta y_6 = x_{out} - x_{in} = \mathscr{A}_1(J - J^{N+1}) : -\Delta y_1 = y_{out} - y_{in} = \mathscr{B}_1(J^* - J^{*N+1})$$

Dividing the concentration differences in pairs to eliminate the first constant $\mathscr{A}_1$ or $\mathscr{B}_1$ gives expressions relating the fourteen separation factors listed in Table 2.1 to N and J or J^*. However, the expressions involving J^* are the inverses of the corresponding expressions involving J. This illustrates how each of the expressions is duplicated by inversion and indicates how they may be reduced to the eight basic formulae listed in Table 2.3.

APPENDIX 2c
EXAMPLES OF THE USE OF FORMULAE RELATING N, J AND Q

Suppose a solute is being extracted from the G phase to the L phase, which have flowrates of 14 and 20 gal/min respectively. The concentrations in g/l are $x_{in} = 0$, $x_{out} = 6{\cdot}6$, $y_{in} = 10$ and $y_{out} = 0{\cdot}57$ and the equilibrium relationship is $y = x$, so $m = 1$ and $c = 0$. The operating diagram is shown in Fig. 3.1.

The value of $J = mG/L = 1 \times 14/20 = 0{\cdot}7$ and of $J^* = 1/J = 1{\cdot}43$. Thus $1 - J = 0{\cdot}3$ and $1 - J^* = -0{\cdot}43$; $\ln 1/J = 0{\cdot}358$ and $\ln 1/J^* = -0{\cdot}358$. The six concentration differences are;

$$\Delta y_1 = y_{in} - y_{out} = 9{\cdot}43$$
$$\Delta y_2 = y_{in} - y(x_{in}) = 10$$
$$\Delta y_3 = y_{in} - y(x_{out}) = 3{\cdot}4$$
$$\Delta y_4 = y_{out} - y(x_{in}) = 0{\cdot}57$$
$$\Delta y_5 = y(x_{out}) - y_{out} = 6{\cdot}03$$
$$\Delta y_6 = y(x_{out}) - y(x_{in}) = 6{\cdot}6$$

TABLE A2c.1. CALCULATION OF SEPARATION FACTORS AND NUMBERS OF STAGES AND TRANSFER UNITS BASED ON THE FORMULAE GIVEN IN TABLE 2.3

Separation factor Q	$(1/J)^{N_S}$ or $\exp(N_G(1-J))$	N_S	N_G	Formula	Inverse separation factor Q^*	$(1/J^*)^{N_S}$ or $\exp(N_L(1-J^*))$	N_S	N_L
$Q_{1/2} = 0{\cdot}943$	$\frac{1-JQ}{1-Q} = 5{\cdot}97$	5·0	5·96	I	$Q_{6/2} = 0{\cdot}66$	$\frac{1-J^*Q^*}{1-Q'} = 0{\cdot}168$	5·0	4·17
$Q_{1/3} = 2{\cdot}77$	$\frac{1}{1+Q(J-1)} = 5{\cdot}97$	5·0	5·96	II	$Q_{6/4} = 11{\cdot}58$	$\frac{1}{1+Q^*(J^*-1)} = 0{\cdot}168$	5·0	4·17
$Q_{1/4} = 16{\cdot}5$	$1 + Q(1-J) = 5{\cdot}97$	5·0	5·96	III	$Q_{6/3} = 1{\cdot}94$	$1 + Q^*(1-J^*) = 0{\cdot}168$	5·0	4·17
$Q_{1/5} = 1{\cdot}56$	$\frac{1-Q}{1-JQ} = 5{\cdot}97$	5·0	5·96	IV	$Q_{6/5} = 1{\cdot}09$	$\frac{1-Q^*}{1-J^*Q^*} = 0{\cdot}168$	5·0	4·17
$Q_{2/4} = 17{\cdot}52$	$(1-J)Q + J = 5{\cdot}97$	5·0	5·96	V	$Q_{2/3} = 2{\cdot}94$	$(1-J^*)Q^* + J' = 0{\cdot}168$	5·0	4·17
$Q_{2/5} = 1{\cdot}66$	$\frac{J-Q}{1-JQ} = 5{\cdot}97$	5·0	5·96	VI	$Q_{2/5} = 1{\cdot}66$	$\frac{J^*-Q^*}{1-J^*Q^*} = 0{\cdot}168$	5·0	4·17
$Q_{3/4} = 5{\cdot}97$	$Q = 5{\cdot}97$	5·0	5·96	VII	$Q_{3/4} = 0{\cdot}173$	$Q^* = 0{\cdot}168$	5·0	4·17
$Q_{5/3} = 1{\cdot}77$	$\frac{1}{J+Q(J-1)} = 5{\cdot}97$	5·0	5·96	VIII	$Q_{5/4} = 10{\cdot}6$	$\frac{1}{J^*+Q^*(J^*-1)} = 0{\cdot}168$	5·0	4·17

These are used to calculate the separation factors and their inverses listed in Table 2.3 the values of which aregiven in Table A2c.1. Using the eight basic formulae listed in Table 2.3 the calculated numbers of stages or transfer units are $N_S = 5{\cdot}0$, $N_G = 5{\cdot}96$ and $N_L = 4{\cdot}17$. The ratio N_L/N_G is, of course, equal to J.

The separation factors may also be used to calculate the numbers of stages and transfer units from formulae suggested by other authors.

For example, one of the formulae suggested by the third edition of the *Chemical Engineers' Handbook*[9] and Sherwood and Pigford[8] is

$$N_S \ln 1/J^* = N_L(1 - J^*) = \ln [(J^* - 1) Q_{5/4} + J^*] \qquad \text{A2c(1)}$$

and the argument of the right-hand side is

$$(J^* - 1) Q_{5/4} + J^* = (1{\cdot}43 - 1) \times 10{\cdot}6 + 1{\cdot}43 = 6{\cdot}00.$$

The formula thus becomes

$$N_S \ln 1/1{\cdot}43 = N_L(-0{\cdot}43) = \ln 6$$

which leads to $N_S = -5$, $N_L = -4{\cdot}17$ and $N_G = N_L/J = -5{\cdot}97$. These are the correct absolute values but have the wrong sign.

The fourth edition of the *Chemical Engineers' Handbook*[7] quotes the same formula for N_S as equation A2c(1) but this time explicit in Q. It also suggests the use of

$$N_S \ln 1/J^* = \ln [(1 - J^*) Q_{5/4} + J^*] \qquad \text{A2c(2)}$$

when the argument of the right-hand side becomes

$$(1 - J^*) Q_{5/4} + J^* = (1 - 1{\cdot}43) \times 10{\cdot}6 + 1{\cdot}43 = -3{\cdot}14$$

which leads to imaginary values of N_S.

APPENDIX 2d
NUMERICAL EXAMPLES OF HEAT TRANSFER USING $N_T(1 - J) = \ln Q$

Example 1

1 lb/sec of oil of specific heat 0·5 B.t.u./lb °F is cooled from 180 °C to 20 °C with air flowing counter-currently in a shell and tube exchanger, with a total tube area of 100 ft². If the air temperature rises from 15 to 75 °C what is the heat transfer coefficient?

Suppose the air is phase G and the oil phase L.
Then

$$Q_{3/4} = \frac{\theta_{G,\,\text{in}} - \theta_{L,\,\text{out}}}{\theta_{G,\,\text{out}} - \theta_{L,\,\text{in}}} = \frac{-5}{-105} = \frac{1}{21} \quad \text{and so} \quad Q_{4/3} = 21$$

$$J = \frac{W_G}{W_L} = \frac{\theta_{L,\,\text{in}} - \theta_{L,\,\text{out}}}{\theta_{G,\,\text{out}} - \theta_{G,\,\text{in}}} = \frac{160}{60} \quad \text{so} \quad J^* = \frac{3}{8}$$

$$W_L = 1 \times 3600 \times 0{\cdot}5 = 1800 \text{ B.t.u./hr °F.}$$

Using

$$N_L(1 - J^*) = \ln Q_{4/3}$$

$$\frac{UA}{W_L}\left(1 - \frac{3}{8}\right) = \ln 21$$

and so

$$U = \frac{1800}{100} \times \frac{8}{5} \ln 21 = 88 \text{ B.t.u./hr ft}^2 \text{ °F.}$$

Example 2

Calculate the area of tube which is required to heat a flow of 4 ft³/min of air (measured at 60 °F and atmospheric pressure) from 60 °F to 500 °F if the tube wall is maintained at 600 °F.

The heat transfer coefficient U calculated from

$$\text{Nu} = 0{\cdot}023 \text{ Re}^{0\cdot8} \text{ Pr}^{0\cdot4}$$

is 14·45 B.t.u./hr ft² °F, the density of air at 60 °F is 0·076 lb/ft³ and the mean specific heat of air in the range 60 °F to about 500 °F is 0·244 B.t.u./lb °F.

Suppose air is the G phase so

$$W_G = Gd_GC_G = 4 \times 60 \times 0{\cdot}076 \times 0{\cdot}244 = 4{\cdot}46 \text{ B.t.u./hr ft}^2 \text{ °F.}$$

As θ_L is constant W_L must be infinite and so $J = W_G/W_L = 0$.
Now

$$Q_{3/4} = \frac{\theta_{G.\,\text{in}} - \theta_L}{\theta_{G,\,\text{out}} - \theta_L} = \frac{-540}{-100} = 5{\cdot}4$$

and so using $N_G = \dfrac{UA}{W_G} = \ln Q_{3/4}$ gives $A = 4{\cdot}46 \ln (5{\cdot}4)/14{\cdot}45 = 0{\cdot}521 \text{ ft}^2$.

CHAPTER 3

Errors in the Calculation of Numbers of Stages and Transfer Units

3.1. Introduction

Experience shows that the graphical or analytical calculation of numbers of stages and transfer units can often lead to large errors; for example, the qualitative effects of a pinch are well known. A glance at

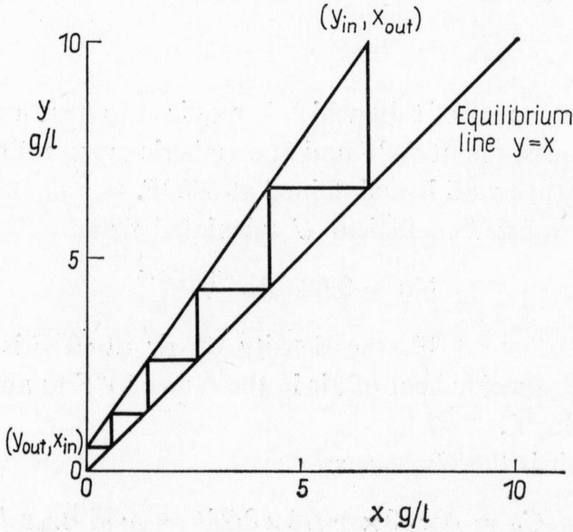

FIG. 3.1. Operating diagram for counter-current extraction with $J = 0{\cdot}7$ and $Q = 6$ (so that $N_S = 5$ and $N_T = 6$).

Fig. 3.1 shows that a small variation in J will lead to a large variation in N and a small variation in y_{out} to a large variation in Q. By making these effects quantitative rather than qualitative some idea of the degree of accuracy or control needed in the fundamental measurements may be obtained. The variations could be due to finite changes in the variables, such as a drop from 10 to 9·5 gal/min in a flowrate, or to errors in the measurements involved, such as an uncertainty of $\pm 0{\cdot}5$ in a flowmeter reading of 10 gal/min.

In the design of a contactor the separation factor Q is usually fixed and the number of stages or transfer units is calculated for a reasonable value of the extraction factor, J. The designer is interested in knowing what effect a variation in J, or in the number of theoretical stages or transfer units N, will have on Q, when the contactor is operated. As Q can vary because the feed concentrations vary, even at constant output concentrations, he is also interested in the effect of small variations in J and Q on the required number of theoretical stages or transfer units. Allowance can then be made for such variations in the design.

Similarly in the evaluation of a contactor the number of theoretical stages or transfer units, N is calculated from experimental values of J and Q. One would like to know the error in N resulting from the experimental errors in J and Q.

In the operation of a contactor the value of N is, broadly speaking, fixed and the desired separation Q is achieved by adjusting J. The operator would thus like to know what effect a variation in J will have on Q. If it is possible to vary the number of theoretical stages or transfer units he is also interested in the effect of variations in N on Q. For a stagewise process in which the number of actual stages N_A is fixed, this is equivalent to the effect of variations in the overall stage efficiency E on Q.

Thus in general the effect of variations and errors in J and Q on N and also of J and N on Q are of interest, but not of N and Q on J.

3.2. Theory

All the equations discussed in Chapter 2 may be written in the general form $N = f(J, Q)$ and so for small changes in the variables[1]

$$\delta N = \left(\frac{\partial N}{\partial J}\right)_Q \delta J + \left(\frac{\partial N}{\partial Q}\right)_J \delta Q \qquad 3(1)$$

and for small errors

$$e_N^2 = \left(\frac{\partial N}{\partial J}\right)_Q^2 e_J^2 + \left(\frac{\partial N}{\partial Q}\right)_J^2 e_Q^2 \qquad 3(2)$$

where e is the standard deviation and e^2 the variance of the variables referred to by the subscripts. For larger changes or errors Taylor series expansions involving higher-order derivatives must be used. According to Davies[(2)] equation 3(2) applies if each standard deviation is less than 20% of the mean value of the variable.

Dividing equations 3(1) and 3(2) by N and N^2 respectively and rearranging, yields relationships between the relative changes or errors (the percentage error is sometimes called the coefficient of variation),

$$\frac{\delta N}{N} = \left(\frac{\partial N}{\partial J}\right)_Q \frac{J}{N}\frac{\delta J}{J} + \left(\frac{\partial N}{\partial Q}\right)_J \frac{Q}{N}\frac{\delta Q}{Q} \qquad 3(3)$$

and

$$\left(\frac{e_N}{N}\right)^2 = \left(\frac{\partial N}{\partial J}\right)_Q^2 \frac{J^2}{N^2}\left(\frac{e_J}{J}\right)^2 + \left(\frac{\partial N}{\partial Q}\right)_J^2 \frac{Q^2}{N^2}\left(\frac{e_Q}{Q}\right)^2 \qquad 3(4)$$

For a stagewise process with N_A actual stages the ratio N_S/N_A is the overall stage efficiency E and so $\delta N/N$ and $(e_N/N)^2$ become $\delta E/E$ and $(e_E/E)^2$ respectively.

The equations of Chapter 2 may be re-expressed in the form $Q = f(N, J)$ and so similar expressions to 3(1), 3(2), 3(3) and 3(4) may be written giving the absolute and fractional changes and errors in Q in terms of those in N and J.

In order to find the effect of variations in J and Q on N, or in N and J on Q the values of the partial derivatives must be determined.

3.3. Discussion

The simplest formula discussed in Chapter 2 was

$$N_S \ln 1/J = N_T(1 - J) = \ln Q \qquad 3(5)$$

where

$$J = \frac{mG}{L}, \quad Q = \frac{y_{\text{in}} - y(x_{\text{out}})}{y_{\text{out}} - y(x_{\text{in}})} \quad \text{and} \quad N_T = N_G$$

or

$$J = \frac{L}{mG}, \quad Q = \frac{y_{\text{out}} - y(x_{\text{in}})}{y_{\text{in}} - y(x_{\text{out}})} \quad \text{and} \quad N_T = N_L$$

and the linear equilibrium relationship is

$$y(x) = mx + c \tag{3(6)}$$

Table 3.1 gives the algebraic expressions for the partial derivatives of equation 3 (5). If Q is defined differently as in Table 2.1 the relationship between N, J and Q is given by Table 2.3 and the resulting

TABLE 3.1. EXPRESSIONS FOR PARTIAL DERIVATIVES AND VALUES WHEN $J = 0{\cdot}7$ AND $Q = 6$ (so $N_S = 5$ and $N_T = 6$)

Derivative	Expression	Value	Derivative	Expression	Value
$\left(\frac{\partial N_S}{\partial J}\right)_Q$	$= \frac{N_S}{J \ln 1/J}$	20	$\left(\frac{\partial N_S}{\partial J}\right)_Q \frac{J}{N_S}$	$= \frac{1}{\ln 1/J}$	28
$\left(\frac{\partial Q}{\partial N_S}\right)_J$	$= Q \ln 1/J$	2·2	$\left(\frac{\partial Q}{\partial N_S}\right)_J \frac{N_S}{Q}$	$= \ln Q$	1·8
$\left(\frac{\partial Q}{\partial J}\right)_{N_S}$	$= \frac{N_S Q}{J}$	−43	$\left(\frac{\partial Q}{\partial J}\right)_{N_S} \frac{J}{Q}$	$= -N_S$	−5
$\left(\frac{\partial N_T}{\partial J}\right)_Q$	$= \frac{N_T}{1 - J}$	20	$\left(\frac{\partial N_T}{\partial J}\right)_Q \frac{J}{N_T}$	$= \frac{J}{1 - J}$	2·3
$\left(\frac{\partial Q}{\partial N_T}\right)_J$	$= Q(1 - J)$	1·8	$\left(\frac{\partial Q}{\partial N_T}\right)_J \frac{N_T}{Q}$	$= \ln Q$	1·8
$\left(\frac{\partial Q}{\partial J}\right)_{N_T}$	$= -N_T Q$	−36	$\left(\frac{\partial Q}{\partial J}\right)_{N_T} \frac{J}{Q}$	$= -N_T J$	−4·2

partial derivatives are more complicated. Because only three variables are involved $(\partial Q/\partial N)_J$ is the reciprocal of $(\partial N/\partial Q)_J$. Inspection shows that in general $(\partial Q/\partial N)_J$, $(\partial Q/\partial J)_N$, $(\partial Q/\partial N)_J(N/Q)$, and $(\partial Q/\partial J)_N(J/Q)$ are large when N and Q are large. This is also true of $(\partial N/\partial J)_Q$ but $(\partial N/\partial J)_Q(J/N)$ is independent of N and Q; however, both are large when J is close to unity. Most of the partial derivatives may be related to N, J and Q by straight lines as shown, for example, by Figs. 3.2 and 3.3 which give the absolute or fractional error in Q resulting from an error in N_S or N_T, as functions of Q or N for different values of J.

Table 3.1 also shows the magnitude of the derivatives when $J = 0{\cdot}7$ and $Q = 6$ (so that $N_S = 5$ and $N_T = 6$). This case is illustrated in Fig. 3.1

with $y_{in} = 10$ (g/l say) and $x_{in} = 0$, which emphasises that the values chosen are very conservative. Had N and Q been larger so that a pinch developed at one end of the operating line the values of the derivatives would have been much greater. (Note the effect for example of putting $J = 0{\cdot}5$ and $Q = 1000$ so that $N_S = 10$ and $N_T = 14$.) In spite of this

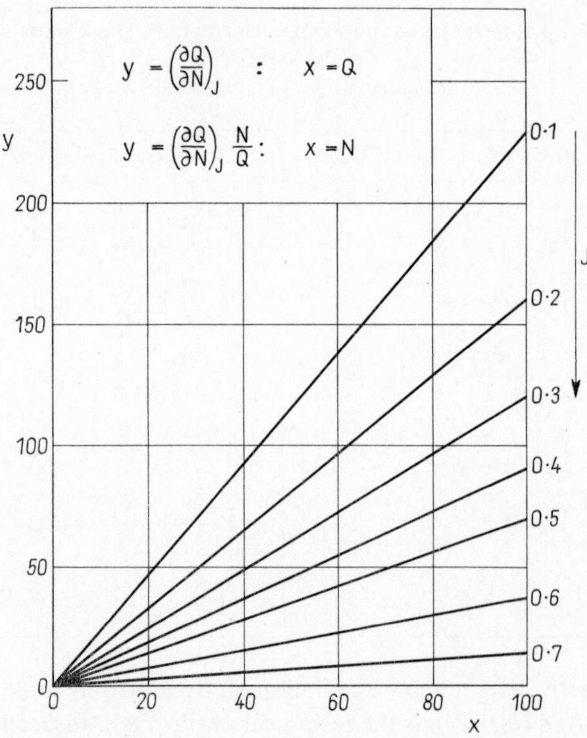

FIG. 3.2. Variation of ratio of absolute or fractional errors in $Q_{3/4}$ and N_S with $Q_{3/4}$ or N_S for different values of J.

the values of the derivatives are substantial and show that even small fractional changes or errors in J and N may be magnified many times.

The variations in J and Q may be calculated from the variations in the experimental measurements of flowrates, end concentrations and equilibrium data as illustrated in Example 3.1.

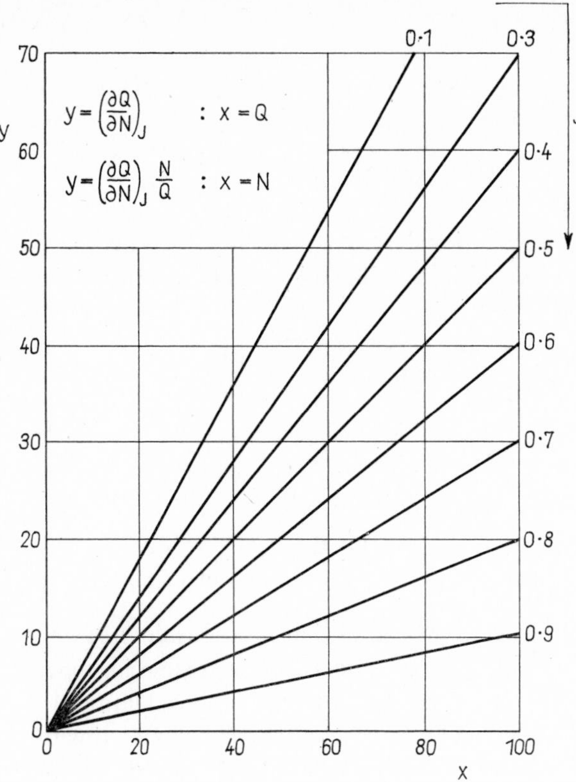

FIG. 3.3. Variation of ratio of absolute or fractional errors in Q_{3J4} and N_T with Q_{3J4} or N_T for different values of J.

Example 3.1

In an experiment to evaluate the stage efficiency of a contactor containing six real stages the measured flowrates are $L = 20$ gal/min, $G = 14$ gal/min and the end concentrations $x_{in} = 0$, $y_{in} = 10$ g/l, $x_{out} = 6{\cdot}6$ g/l and $y_{out} = 0{\cdot}57$ g/l. The flowrates may be estimated to ± 1 gal/min and the concentrations to $\pm 0{\cdot}05$ g/l (there is no error in x_{in}). The equilibrium line is $y = x$ and there is a 1% error in the slope. What is the error in the calculated value of the stage efficiency?

The quantities in this example are illustrated in Fig. 3.1.

SOLUTION

$$J = \frac{mG}{L} = \frac{1 \times 14}{20} = 0{\cdot}7$$

and

$$Q = \frac{y_{\text{in}} - y(x_{\text{out}})}{y_{\text{out}} - y(x_{\text{in}})} = \frac{10 - 6{\cdot}6}{0{\cdot}57 - 0} = 6{\cdot}0$$

and so from equation 3(5), $N_S = 5$ and $E = N_S/N_A = 5/6 = 83\%$.

Using the definitions of J and Q in equation 3(5) their variances are given respectively by

$$\frac{e_J^2}{J^2} = \frac{e_m^2}{m^2} + \frac{e_G^2}{G^2} + \frac{e_L^2}{L^2} \qquad 3(7)$$

and

$$\frac{e_Q^2}{Q^2} = \frac{e_{y_{\text{in}}}^2 + e_{y(x_{\text{out}})}^2}{(y_{\text{in}} - y(x_{\text{out}}))^2} + \frac{e_{y_{\text{out}}}^2 + e^2{}_{y(x_{\text{in}})}}{(y_{\text{out}} - y(x_{\text{in}}))^2} \qquad 3(8)$$

where

$$\frac{e_{y(x)}^2}{(y(x))^2} = \frac{e_m^2}{m^2} + \frac{e_x^2}{x^2} \qquad 3(9)$$

which follows from the equilibrium relationship 3(6) when $c = 0$. The standard deviation of a quantity is obtained by taking repeated measurements which are independent of the other quantities.[(2)] As pointed out by Rowe[(3)] and Davies[(4)] the experimenter's estimate of the uncertainty in a measurement may be taken as the standard deviation for most engineering purposes. Thus using equation 3(7),

$$e_J^2/J^2 = 1 \times 10^{-4} + 1/14^2 + 1/20^2 = 77 \times 10^{-4}$$

and so there is a 9% error in J almost completely due to the flowmeter errors. Using equation 3(9)

$$e_{y(x_{\text{out}})}^2 = 6{\cdot}6^2 \times 1 \times 10^{-4} + 25 \times 10^{-4} = 69 \times 10^{-4}$$

and $\quad e_{y(x_{\text{in}})} = 0 \quad \text{as} \quad y(x_{\text{in}}) = 0.$

From equation 3(8)

$$e_Q^2/Q^2 = (25 \times 10^{-4} + 69 \times 10^{-4})/(3{\cdot}4)^2 + 25 \times 10^{-4}/0{\cdot}57^2$$
$$= 86{\cdot}2 \times 10^{-4}$$

so there is an 11% error in Q mainly due to the 9% error in y_{out}.

Table 3.1 shows that $(\partial N_S/\partial J)_Q(J/N_S) = 2{\cdot}8$ and $(\partial N_S/\partial Q)_J(Q/N_S) = 0{\cdot}56$ so it follows from equation 3(4) that $e_{N_S}^2/N_S^2 = 2{\cdot}8^2 \times 77 \times 10^{-4} + 0{\cdot}56^2 \times 86{\cdot}2 \times 10^{-4} = 629 \times 10^{-4}$. Thus e_{N_S}/N_S or e_E/E equals 0·251 and so $N_S = 5{\cdot}0 \pm 1{\cdot}3$ and $E = 83 \pm 21\%$.

Practically all the uncertainty is due to the magnification of the error in J and so to obtain a better result it would appear the flowrates would have to be measured more accurately. Even though the relative error in Q is greater than that in J it has little effect on E and so there is no point in trying to reduce the large relative error in y_{out}.

However, if J is estimated from the overall balance

$$J(y_{in} - y_{out}) = y(x_{out}) \tag{3(10)}$$

its variance is then given by

$$\frac{e_J^2}{J^2} = \frac{e_{y(x_{out})}^2}{(y(x_{out}))^2} + \frac{e_{y_{in}}^2 + e_{y_{out}}^2}{(y_{in} - y_{out})^2} \tag{3(11)}$$

and so $e_J^2/J^2 = 69 \times 10^{-4}/(6{\cdot}6)^2 + 50 \times 10^{-4}/(9{\cdot}4)^2 = 2{\cdot}16 \times 10^{-4}$ which is much lower than the previous estimate.

It follows from equation 3(6) that the variance in N_S is given by $e_{N_S}^2/N^2 = 2{\cdot}8^2 \times 2{\cdot}16 \times 10^{-4} + 0{\cdot}56^2 \times 86{\cdot}2 \times 10^{-4} = 43{\cdot}9 \times 10^{-4}$ and so $e_{N_S}/N_S = e_E/E = 0{\cdot}066$. The value of J estimated from equation 3(10) is 0·7 as before and so $N_S = 5{\cdot}0 \pm 0{\cdot}3$ and $E = 83 \pm 6\%$. (In general because the mean values of the flowrates and end concentrations will not be self consistent, the two estimates of J, and hence of N_S and E, will be different, however.)

The errors in N_S and E are now much smaller than before and the contribution due to the errors in J and Q about the same. Consideration of the arithmetical values in equations 3(10) and 3(8) shows that the errors in J and Q are mainly due to the errors in $y(x_{out})$ and y_{out} respectively. The errors in N_S and E may thus be reduced further by reducing the errors in y_{out}, x_{out} and m.

Thus by careful consideration of the effect of errors in the experimental measurements it is possible to obtain the best estimate of the overall stage efficiency from the available data and to see where it is most worth while reducing the errors.

Notation for Chapter 3

c	intercept of equilibrium line on y-axis
e	standard deviation
e^2	variance
E	overall stage efficiency
G	flowrate of phase in which solute concentration is y
J	extraction factor ($J = mG/L$)
L	flowrate of phase in which solute concentration is x
m	slope of equilibrium line (equation $y = mx + c$)
N	number of theoretical stages or transfer units
N_A	number of actual stages
N_G	number of overall transfer units based on phase of flowrate G
N_L	number of overall transfer units based on phase of flowrate L
N_S	number of theoretical stages
N_T	number of transfer units
Q	separation factor
x	solute concentration in phase of flowrate L
x_{in}	x concentration entering contactor
x_{out}	x concentration leaving contactor
y	solute concentration in phase of flowrate G
$y(x)$	y concentration in equilibrium with concentration x ($y(x) = mx + c$)
y_{in}	y concentration entering contactor
y_{out}	y concentration leaving contactor

The above quantities may be expressed in any set of consistent units in which force and mass are not defined independently.

References

1. W. E. Milne, *Numerical Calculus*, Princeton University Press, 1939.
2. O. L. Davies, *Statistical Methods in Research and Production*, 3rd ed., Edinburgh; Oliver & Boyd, 1957.
3. P. N. Rowe, *The Chemical Engineer*, C.E.69 (1963).
4. O. L. Davies, *The Design and Analysis of Industrial Experiments*, Edinburgh; Oliver & Boyd, 1954.

CHAPTER 4

The Optimisation of Counter-current Forward and Back Extraction

4.1. Introduction

Forward extraction is used when two solutes must be separated by extracting one into a solvent. This solvent is chosen mainly for its selectivity[(1)] and extracting power and so it may be necessary to back extract the solute into another solvent before the next stage in the process. For example, the intermediate solvent may be expensive and back extraction of the solute into a cheaper solvent allows the expensive solvent to be recirculated. Alternatively the intermediate solvent may be difficult to handle; it may be toxic, flammable, corrosive, volatile or viscous. Recirculation allows the difficult solvent to be contained. Again it may not be possible to obtain the pure solute by separation from the intermediate solvent; a further solvent must be used from which the solute can be separated.

Forward and back extraction is widely used in gas absorption[(2)] and solvent extraction.[(3)] Some particular examples which illustrate the variety of its application are listed in Table 4.1.

A typical arrangement for forward and back extraction is shown in Fig. 4.1; L is the flowrate of the initial solvent, S the flowrate of the intermediate solvent and G the flowrate of the final solvent. These will be referred to as the L, S and G phases respectively. The initial and intermediate solvents flow counter-currently in contactor 1 and the intermediate and final solvents counter-currently in contactor 2. The contactors may be differential or stagewise in operation. The concentrations x, y and z refer to the solute which is being extracted in the L, G and S phases respectively. The term forward and back extraction is used as other terms such as absorption and stripping have such a

TABLE 4.1. EXAMPLES OF FORWARD AND BACK EXTRACTION

Solute	Solvents			Application	Reference
	Initial	Intermediate	Final		
H_2S	Coke oven gas	Dilute aqueous Na_2CO_3	Air	Seaboard Process	2
Glycerides	Soybean oil	Furfural	Naphtha	Purification	4, 5
Phenols	Gas liquor	Light oil	Aqueous caustic soda	Recovery	3
Uranium or Plutonium	Aqueous HNO_3	20% Tributyl phosphate in odourless kerosene (TBP/OK)	Water	Purex Process. Typical of many processes in nuclear fuel reprocessing	6–9
Uranium	Sulphate leach liquor	Amine in kerosene	Aqueous Na_2CO_3	Typical of many processes in ore extraction	10–15
Mercaptans	Gasolene	Aqueous caustic solutions	Steam	Sweeting of sour gasolene	16, 17
H_2S, CO_2	Acid gas	Ethanolamine	Steam	Girbitol Process	2
Trace impurities	Lubricating oil	β,β' dichloroethyl ether	Steam	Chlorex Process	18
Bromine	Sea water	Air	Aqueous Na_2CO_3	Ethyl Dow Process	19

variety of meaning. For example, both forward and back extraction are sometimes referred to as stripping.

A typical operating diagram for forward and back extraction is shown in Fig. 4.2. Transfer is visualised as being from the x to y phases but it could of course be in the opposite direction. In general the aim

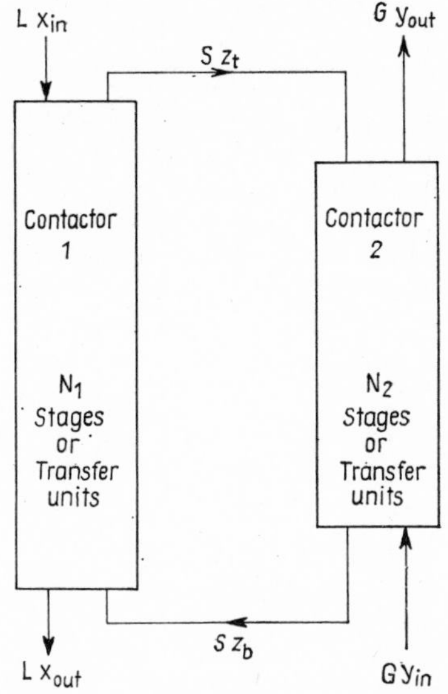

FIG. 4.1. Flows and concentrations in counter-current forward and back extraction.

will be to extract as much as possible of the solute from the first solvent into the final solvent; x_{out} will be specified at some low value and y_{out} at some high value.

If L and G are fixed, good extraction, and hence a low value of x_{out}, may be obtained in the first contactor by keeping the value of z everywhere low. Good extraction, and hence a high value of y_{out}, may be similarly obtained in the second contactor by keeping the value of z everywhere high. If the level of z in the system as a whole is raised, more stages or transfer units will be required in the first contactor and less in the second contactor. The opposite will apply if the level of z is lowered. An optimum level of z thus exists for which the total number of stages or transfer units in both contactors is a minimum when the separation is fixed. Alternatively, for a fixed number of stages or transfer units there is an optimum level of z for which the separation is a maximum. The level of z obviously depends on the recirculation rate of the intermediate solvent, S.

When operating a given forward and back extraction system one would like to know the optimum value of S, for which the separation

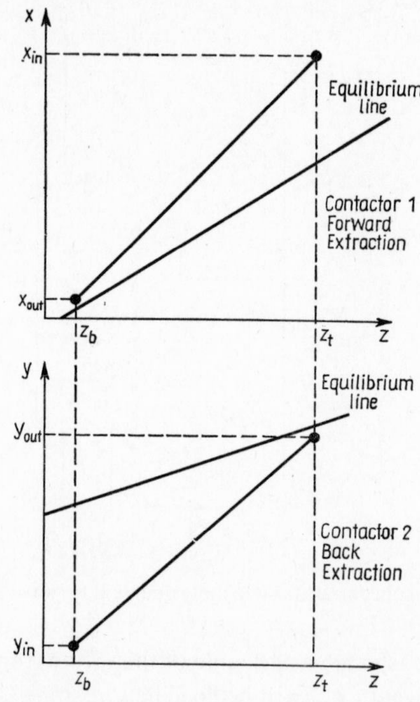

FIG. 4.2. Operating diagram for forward and back extraction.

is a maximum. When designing a forward and back extraction system one would like to know how to distribute the number of stages or transfer units between the two contactors so that the total number of stages or transfer units is a minimum for the given separation.

Simple solutions are presented to these two problems assuming linear equilibrium data and immiscible phases so that the equilibrium and operating lines are straight. Such solutions will be necessarily restricted in application but they enable quick and relatively easy answers to be obtained which may be satisfactory for preliminary design.

The result could provide a reasonable start to a more precise and complicated method, thus limiting the more tedious work involved or saving computer time.

A complete optimisation would be based on the capital and operating costs. In design, minimising the number of stages or transfer units may go some way towards minimising the capital cost, but not necessarily the operating costs. In operation one must balance the separation achieved against the cost of recirculation of the intermediate solvent. The economics of a forward and back extraction system is discussed by Treybal(20), Sherwood and Pigford(21) and Colburn(22). Such an analysis is necessarily complicated and involves a number of arbitrary assumptions. It is still useful to have relatively simple expressions which are exact within the limitations specified.

Aris, Rudd and Amundson(23) in their paper on optimum cross-current extraction state: "For processes with many operating variables the standard method of setting the derivatives of the profit with respect to all the variables simultaneously equal to zero rapidly leads to equations of intolerable complexity. Direct numerical calculations in search of the maximum are laborious even when guided by a gradient technique." These authors apply the notion of dynamic programming to the problem of allocating an extracting solvent to the various stages of a cross-current extraction unit. Rudd and Blum(24) extend the analysis to include recycle. Aris(25), again using dynamic programming, shows how to optimise the operating conditions in a cascade of stirred tank reactors with and without bypassing of the feed. Jenson and Jeffries(26) give a detailed cost analysis of counter-current extraction and maximise the profit function with respect to the number of stages and flow-rate of the extracting solvent. Srygley and Holland(27) discuss the optimum design of conventional and complex distillation columns. There are many more papers (e.g. references 28 to 42) which discuss the different methods of optimisation and their application to chemical processes.

4.2. Equations Governing the Optimisation of Forward and Back Extraction

Basic Equations

The flows and concentrations are given in Fig. 4.1. When there is a linear equilibrium relationship

$$z = m_1 x + c_1 \tag{4(1)}$$

the number of stages, N_{S1} or overall transfer units based on the L phase, N_{L1} in contactor 1 is given by equations 2(10) and (11) as

$$N_{S1} \ln \frac{1}{J_1} = \frac{N_{L1}}{J_1}(1 - J_1) = \ln Q_1 \tag{4(2)}$$

where J_1 is an extraction factor given by

$$J_1 = \frac{m_1 S}{L} \tag{4(3)}$$

and Q_1 is a separation factor given by

$$Q_1 = \frac{z_b - (m_1 x_{\text{out}} + c_1)}{z_t - (m_1 x_{\text{in}} + c_1)} \tag{4(4)}$$

When there is a linear equilibrium relationship

$$y = m_2 z + c_2 \tag{4(5)}$$

the number of stages N_{S2} or overall transfer units based on the G phase, N_{G2} in contactor 2 is given by

$$N_{S2} \ln \frac{1}{J_2} = N_{G2}(1 - J_2) = \ln Q_2 \tag{4(6)}$$

where

$$J_2 = \frac{m_2 G}{S} \tag{4(7)}$$

and

$$Q_2 = \frac{y_{\text{in}} - (m_2 z_b + c_2)}{y_{\text{out}} - (m_2 z_t + c_2)} \tag{4(8)}$$

N_{L1} and N_{G2} are chosen as the numbers of transfer units in contactors 1 and 2 respectively because they will be constant if L and G are constant even if S varies. This is not true of the numbers of overall transfer units based on the S phase in contactors 1 and 2.

Overall Separation Factor

In equations 4(4) and (8), for example, $(m_1 x_{in} + c_1)$ is the z concentration in equilibrium with x_{in} and $(m_2 z_t + c_2)$ is the y concentration in equilibrium with z_t. An overall separation factor may thus be defined as

$$Q_o = \frac{y_{in} - (m_1 m_2 x_{out} + m_2 c_1 + c_2)}{y_{out} - (m_1 m_2 x_{in} + m_2 c_1 + c_2)} \qquad 4(9)$$

where, for example, $(m_1 m_2 x_{in} + m_2 c_1 + c_2)$ is the y concentration in equilibrium with z_t and this is in equilibrium with x_{in}. If y_{out} is to be a maximum or a minimum it must approach this equilibrium value. The modulus of the denominator of equation 4(9) must, therefore, be a minimum and hence Q_o must be a maximum.

Equilibrium is always approached at the outlet of the stream with the lower capacity and so it follows that y_{out} will be a maximum in stripping and a minimum in absorption if $J_1 J_2 (= m_1 m_2 G/L)$ is less than one. (Stripping is defined as transfer from the L to G phase and absorption as transfer from the G to L phase.) Similarly if x_{out} is to be a maximum or minimum, Q_o must be a minimum. If $J_1 J_2$ is greater than one, x_{out} will be a maximum in absorption and a minimum in stripping. As shown later, the inverse values of $J_1 J_2$ and Q (denoted *) are $1/J_1 J_2$ and $1/Q_o$ so if we restrict attention to the region $J_1 J_2$ or $(J_1 J_2)^*$ less than one, Q_o or Q_o^* will always be a maximum when y_{out} and x_{out} approach their equilibrium values.

The overall separation factor Q_o is related to the two individual separation factors Q_1 and Q_2 by

$$\frac{1 - J_1 J_2}{Q_o - 1} = \frac{1 - J_2}{Q_2 - 1} + \frac{J_2 - J_1 J_2}{Q_1 - 1} \qquad 4(10)$$

as shown in Appendix 4a.

When $y_{in} = 0$ and $c_1 = c_2 = 0$, Happel, Cauley and Kelly,[16] using the formulae of Yabroff, White and Caselli,[43] derive for a stagewise process

$$\frac{x_{\text{out}}}{x_{\text{in}}} = \frac{(1 - J_2)(J_1^{N_1} - 1) + (J_1 - 1)(J_2 - J_2^{N_2+1})}{(1 - J_2)(J_1^{N_1} - 1) + (J_1^{N_1+1} - 1)(J_2 - J_2^{N_2+1})} \qquad 4(11)$$

This is a particular case of equation 4(10) as shown in Appendix 4b.

Optimisation

The problem is to find N_1/N_2 and the solvent recirculation rate S, such that the overall separation Q_o is a maximum, knowing the operating conditions L, G, m_1, m_2, c_1, c_2 and the total number of stages or transfer units $N_1 + N_2$.

It is thus convenient to write

$$J_1 J_2 = m_1 m_2 G/L = K \qquad 4(12)$$

and

$$N_1 + N_2 = P \qquad 4(13)$$

where K and P are constants.

Thus using equations 4(10), (2), (6), (12) and (13)

$$Q_o = f(N_1, J_1)$$

and so

$$\delta Q_o = \left(\frac{\partial Q_o}{\partial N_1}\right)_{J_1} \delta N_1 + \left(\frac{\partial Q_o}{\partial J_1}\right)_{N_1} \delta J_1 \qquad 4(14)$$

When Q_o is a maximum $\delta Q_o = 0$ and it follows that

$$\left(\frac{\partial Q_o}{\partial N_1}\right)_{J_1} = 0 \quad \text{and} \quad \left(\frac{\partial Q_o}{\partial J_1}\right)_{N_1} = 0.$$

In addition

$$\left(\frac{\partial^2 Q_o}{\partial N_1^2}\right)_{J_1} < 0 \quad \text{and} \quad \left(\frac{\partial^2 Q_o}{\partial J_1^2}\right)_{N_1} < 0$$

and
$$\left(\frac{\partial^2 Q_o}{\partial N_1^2}\right)_{J_1}\left(\frac{\partial^2 Q_o}{\partial J_1^2}\right)_{N_1} > \left(\frac{\partial^2 Q_o}{\partial N_1 \partial J_1}\right)^2$$

as discussed by Mickley, Sherwood and Reid[(44)]. Happel[(45)] gives a classic example of this treatment for the operation of a three stage compressor.

Using equations 4(10), (2), (6) and (13) to obtain $(\partial Q_o/\partial N_1)_{J_1}$ as shown in Appendix 4d and setting this equal to zero gives

$$\frac{(1 - J_1)\, Q_1}{(Q_1 - 1)^2} \ln 1/J_1 = \frac{(1 - J_2)\, Q_2}{J_2\,(Q_2 - 1)^2} \ln 1/J_2 \qquad 4(15)$$

for the stagewise case, and

$$\frac{(1 - J_1)^2\, Q_1}{J_1\,(Q_1 - 1)^2} = \frac{(1 - J_2)^2\, Q_2}{J_2\,(Q_2 - 1)^2} \qquad 4(16)$$

for the differential case.

Using equations 4(10), (2), (6) and (12) to obtain $(\partial Q_o)/(\partial J_1)_{N_1}$ as shown in Appendix 4d and setting this equal to zero gives

$$\frac{Q_1 - 1 - N_1(1 - J_1)\, Q_1}{(Q_1 - 1)^2} = \frac{Q_2 - 1 - N_2(1 - J_2)\, Q_2/J_2}{(Q_2 - 1)^2} \qquad 4(17)$$

for the stagewise case, and

$$\frac{Q_1 - 1 - N_1(1 - J_1)\, Q_1/J_1}{(Q_1 - 1)^2} = \frac{Q_2 - 1 - N_2(1 - J_2)\, Q_2}{(Q_2 - 1)^2} \qquad 4(18)$$

for the differential case.

An alternative way of looking at the problem is to find N_1/N_2 and S such that the total number of stages or transfer units P is a minimum, knowing L, G, m_1, m_2, c_1, c_2 and the separation Q_o. This leads to equations identical with 4(15), (16), (17) and (18) as shown in Appendix 4g.

There are thus nine variables J_1, J_2, N_1, N_2, Q_1, Q_2, Q_o, K and P and seven equations relating these 4(2), (6), (10), (12), (13), (15) or (16), and (17) or (18). If any two variables are fixed the other seven may be determined.

Inversion

The flow diagram in Fig. 4.1 may be physically inverted so that contactors 1 and 2 are interchanged and L, G, S, x_{in}, x_{out}, y_{in}, y_{out}, z_t and z_b are replaced by G, L, S, y_{in}, y_{out}, x_{in}, x_{out}, z_b and z_t respectively. Inversion is denoted by *. The equilibrium relationships in contactors 1 and 2 given by equations 4(1) and (5) may be written

$$x = z/m_1 - c_1/m_1 \quad 4(1a)$$

and

$$z = y/m_2 - c_2/m_2 \quad 4(5a)$$

respectively and so replacing x, y and z by their inverses y, x and z, it follows that m_1, m_2, c_1, and c_2 may be replaced by $1/m_2$, $1/m_1$, $-c_2/m_2$, $-c_1/m_1$ respectively. Thus $J_1 = m_1S/L$ may be replaced by $(1/m_2)\, S/G = 1/J_2$ and the inversion of J_2 is similarly $1/J_1$.

The respective inversions of Q_1, Q_2 and Q_o defined by equations 4(4), (8) and (9) are

$$Q_1^* = \frac{z_t - y_{out}/m_2 + c_2/m_2}{z_b - y_{in}/m_2 + c_2/m_2} = \frac{1}{Q_2} \quad 4(4a)$$

$$Q_2^* = \frac{x_{in} - z_t/m_1 + c_1/m_1}{x_{out} - z_b/m_1 + c_1/m_1} = \frac{1}{Q_1} \quad 4(8a)$$

and

$$Q_o^* = \frac{x_{in} - (y_{out}/m_1m_2 - c_2/m_1m_2 - c_1/m_1)}{x_{out} - (y_{in}/m_1m_2 - c_2/m_1m_2 - c_1/m_1)} = \frac{1}{Q_o} \quad 4(9a)$$

Thus replacing J_1 and Q_1 by $1/J_2$ and $1/Q_2$ in equation 4(2) and comparing the result with equation 4(6) shows that $N_{S1}^* = N_{S2}$ and $N_{L1}^* = N_{G2}$. Similarly inversion of J_2 and Q_2 in equation 4(6) and comparison with equation 4(2) shows that $N_{S2}^* = N_{S1}$ and $N_{G2}^* = N_{L1}$.

We may thus replace J_1, J_2, Q_1, Q_2, N_1, N_2 and Q_o by $1/J_2$, $1/J_1$, $1/Q_2$, $1/Q_1$, N_2, N_1 and $1/Q_o$ respectively in all the stagewise and differential formulae derived for forward and back extraction. This means that only the region J_1J_2 less than one need be considered for if J_1J_2 is greater than one we may invert all the quantities so that $J_1^*J_2^* = 1/J_2J_1$ becomes less than one. As discussed earlier, when J_1J_2 is less than one we require Q_o to be a maximum and when J_1J_2 is greater than one we require $1/Q_o$ to be a maximum. There is thus no inconsistency.

Second-order Derivatives

These are obtained in Appendices 4e and 4f.

For the stagewise case,

$$\left(\frac{\partial^2 Q_o}{\partial N_1^2}\right)_{J_1} = -\left(\frac{Q_o - 1}{Q_2 - 1}\right)^2 \frac{(1 - J_2)\, Q_2 \ln 1/J_2}{1 - J_1 J_2} \times$$

$$\times \left(\frac{Q_1 + 1}{Q_1 - 1} \ln 1/J_1 + \frac{Q_2 + 1}{Q_2 - 1} \ln 1/J_2\right) \quad 4(19)$$

$$\left(\frac{\partial^2 Q_o}{\partial J_1^2}\right)_{N_1} = \left(\frac{Q_o - 1}{Q_1 - 1}\right)^2 \frac{J_2}{J_1^2} \frac{N_1 Q_1}{1 - J_1 J_2} \left(1 + J_1 - N_1(1 - J_1)\frac{Q_1 + 1}{Q_1 - 1}\right)$$

$$+ \left(\frac{Q_o - 1}{Q_2 - 1}\right)^2 \frac{1}{J_1^2} \frac{N_2 Q_2}{1 - J_1 J_2} \left(1 + J_2 - N_2(1 - J_2)\frac{Q_2 + 1}{Q_2 - 1}\right)$$

and 4(20)

$$\frac{\partial^2 Q_o}{\partial N_1 \, \partial J_1} = \left(\frac{Q_o - 1}{Q_1 - 1}\right)^2 \frac{J_2}{J_1} \frac{Q_1}{1 - J_1 J_2} \times$$

$$\times \left(N_1(1 - J_1)\frac{Q_1 + 1}{Q_1 - 1} \ln 1/J_1 - \ln 1/J_1 - (1 - J_1)\right)$$

$$+ \left(\frac{Q_o - 1}{Q_2 - 1}\right)^2 \frac{J_2}{J_1} \frac{Q_2}{1 - J_1 J_2} \times$$

$$\times \left(N_2 \frac{(1 - J_2)}{J_2} \frac{Q_2 + 1}{Q_2 - 1} \ln 1/J_2 - \ln 1/J_2 - \frac{(1 - J_2)}{J_2}\right)$$

4(21)

For the differential case,

$$\left(\frac{\partial^2 Q_o}{\partial N_1^2}\right)_{J_1} = -\left(\frac{Q_o - 1}{Q_2 - 1}\right)^2 \frac{Q_2 (1 - J_2)^2}{1 - J_1 J_2} \times$$

$$\times \left(\frac{Q_1 + 1}{Q_1 - 1} \frac{1 - J_1}{J_1} + \frac{Q_2 + 1}{Q_2 - 1} (1 - J_2)\right) \quad 4(22)$$

$$\left(\frac{\partial^2 Q_o}{\partial J_1^2}\right)_{N_2} = \left(\frac{Q_o - 1}{Q_1 - 1}\right)^2 \frac{J_2}{J_1^3} \frac{N_1 Q_1}{1 - J_1 J_2} \left(2 - N_1 \frac{(1 - J_1)}{J_1} \frac{Q_1 + 1}{Q_1 - 1}\right)$$

$$+ \left(\frac{Q_o - 1}{Q_2 - 1}\right)^2 \frac{J_2^2}{J_1^2} \frac{N_2 Q_2}{1 - J_1 J_2} \left(2 - N_2 (1 - J_2) \frac{Q_2 + 1}{Q_2 - 1}\right) \quad 4(23)$$

and

$$\frac{\partial^2 Q_o}{\partial N_1\,\partial J_1} = \frac{(Q_o - 1)^2}{(Q_1 - 1)}\,\frac{J_2}{J_1^2}\,\frac{(1 - J_1)\,Q_1}{1 - J_1 J_2}\left(N_1\,\frac{(1 - J_1)}{J_1}\,\frac{Q_1 + 1}{Q_1 - 1} - 2\right)$$

$$+ \left(\frac{Q_o - 1}{Q_2 - 1}\right)^2 \frac{J_2}{J_1}\,\frac{(1 - J_2)\,Q_2}{1 - J_1 J_2}\left(N_2\,(1 - J_2)\,\frac{Q_2 + 1}{Q_2 - 1} - 2\right) \quad 4(24)$$

4.3. Solution of the Equations

Differential Case

The differential case is governed by equations 4(2), (6), (10), (12), (13), (16) and (18). Inspection of equations 4(2), (6), (16) and (18) shows these are satisfied by $N_{G2} = N_{L1}/J_1$, $J_1 = J_2$ and $Q_1 = Q_2$. Substitution in equation 4(10) extends this to $Q_1 = Q_2 = Q_O$ and we may thus rewrite equations 4(2) and (6) in the form

$$N_{TO}\,(1 - J_O) = \ln Q_O \quad 4(25)$$

where $N_{TO} = N_{G2} = N_{L1}/J$ and $J_O = J_1 = J_2 = \sqrt{(J_1 J_2)}$. The relationship between N_{TO}, J_O and Q_O is shown in Fig. 4.3.

Equation 4(25) is identical in form with equations 4(2) and (6) which apply to a single contactor and Fig. 4.3 is identical to Fig. 2.2 for a single contactor. A disadvantage of equation 4(25) is that the value of N_{TO} is indeterminate when $J_O = 1$ (for then $Q_O = 1$).

However, in the notation of Chapter 2, Q_O defined by equation 4(9) may be written

$$Q_{3/4} = \frac{y_{\text{in}} - y(x_{\text{out}})}{y_{\text{out}} - y(x_{\text{in}})} \quad 4(26)$$

and $Q_{3/4}$ is related to the other forms of separation factor listed in Table 2.1 through the overall balance which for a forward and back extractor is

$$y(x_{\text{in}}) - y(x_{\text{out}}) = J_1 J_2 (y_{\text{out}} - y_{\text{in}}). \quad 4(27)$$

This is identical with that given by equation 1(9) if J is replaced by $J_1 J_2$. Equation 4(25) may thus be re-expressed in any of the eight basic forms of Table 2.3 if J is replaced by $J_1 J_2$. For example,

$$N_{TO}\,(1 - J_O) = \ln\left[(1 - J_1 J_2)\,Q_{2/4} + J_1 J_2\right] \quad 4(28)$$

where

$$Q_{2/4} = \frac{y_{\text{in}} - y(x_{\text{in}})}{y_{\text{out}} - y(x_{\text{in}})} \quad 4(29)$$

When $J_1 J_2 = 1$ this becomes:

$$N_{TO} = 2\,(Q_{2/4} - 1). \qquad 4(30)$$

Equation 4(28) is analogous to that derived by Colburn[46] for a single contactor.

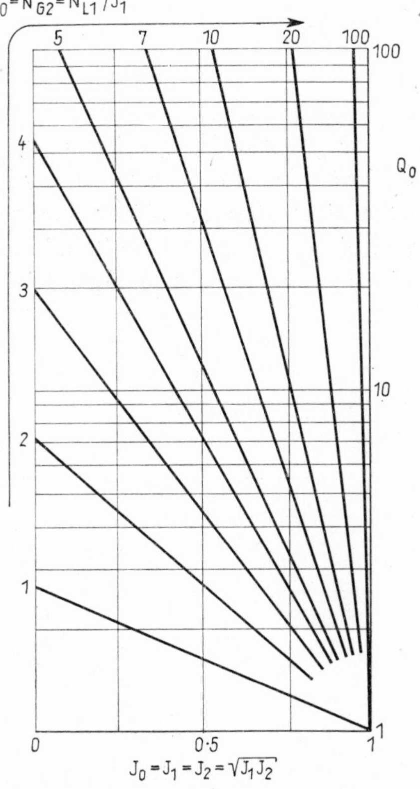

FIG. 4.3. Optimisation of differential, counter-current, forward and back extraction. Variation of maximum overall separation Q_O with overall operating conditions $J_1 J_2$ for different numbers of transfer units N_{TO}.

In design we know the overall separation Q_O and overall operating conditions $J_1 J_2$ $(= m_1 m_2\, G/L)$. The latter gives J_O and N_{TO} follows from equation 4(25). The value of N_{TO} corresponds to the minimum total number of transfer units based on the initial and final solvents $(N_{L1} + N_{G2})$ and $J_O = m_1 S/L = m_2 G/S$ gives the optimum flowrate of the intermediate solvent S. In operation the flowrates L and G of the initial and final solvents must be adjusted so that the values of N_{L1}, N_{G2} and J_1J_2 are related by $N_{G2} = N_{L1}/J$. The maximum separation Q_O then follows from equation 4(25) and the value of J_O gives the optimum value of S. The result will, therefore, be probably more useful in design.

Stagewise Case

The stagewise case is governed by equations 4(2), (6), (10), (12), (13), (15) and (17) and inspection does not reveal any immediate general solution as for the differential case. They were therefore solved as described in Appendix 4h using both an iterative Newton–Raphson technique[47] and a random search technique. Digital computer pro-

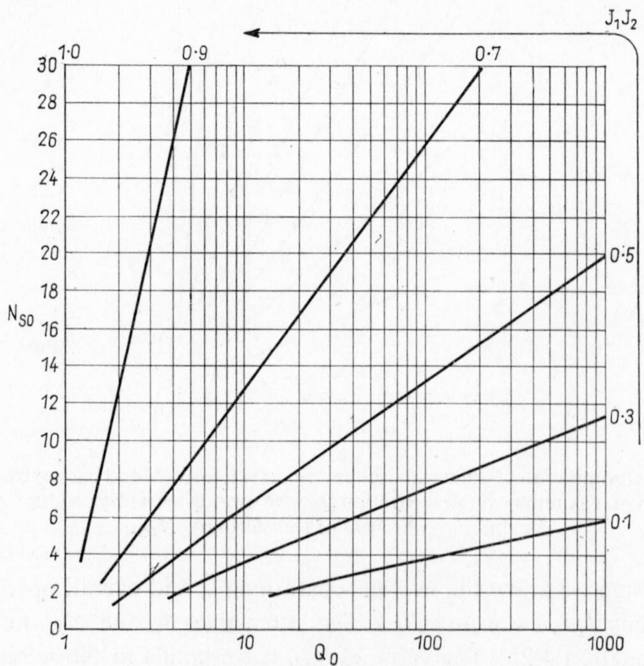

FIG. 4.4. Optimisation of stagewise, counter-current, forward and back extraction. Variation of number of stages N_{SO} with overall separation Q_O for different overall operating conditions J_1J_2.

grams employing these respective techniques are described in Appendices 4m and 4n. The results indicate that N_{S1} and N_{S2} are very nearly equal but that J_1 does not equal J_2. The fractional difference between N_{S1} and N_{S2} decreases as their values increase and as J_1J_2 approaches one. Figure 4.4 is a plot of the mean value N_{SO} versus log Q_O at different values of J_1J_2. The value of N_{S1} is everywhere less than that of N_{S2} but both are within 0·5% of the indicated value of N_{SO}, the maximum error occurring when $J_1J_2 = 0{\cdot}1$ and N_{SO} approaches zero. The form of Fig. 4.4 is suggested by the differential result, given by equation 4(25).

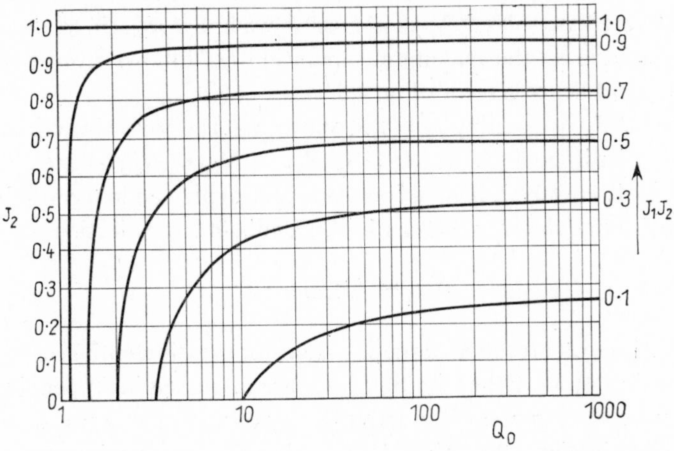

FIG. 4.5. Optimisation of stagewise, counter-current, forward and back extraction. Variation of individual extraction factor J_2 with overall separation Q_O for different overall operating conditions J_1J_2.

Because the computer programs satisfy the physical condition of constant J_1J_2 it is not easy to obtain a stagewise plot analogous to Fig. 4.3. In Fig. 4.4 the lines of constant J_1J_2 correspond closely to

$$N_{SO} \ln 1/\sqrt{(J_1J_2)} = \ln Q_O \qquad 4(31)$$

except when N_{SO} is small and the relationship becomes more exact as the value of J_1J_2 approaches one.

Figure 4.5 gives a plot of J_2 against log Q_O for different values of J_1J_2. This shows that the value of J_1 approaches that of J_2 (as for the differential case) when the value of Q_O is large or when J_1J_2 approaches one.

The applicability of equation 4(31) and of the approximation $J_1 = J_2 = J_1J_2$ may be obtained by comparison with Figs. 4.4 and 4.5. Broadly speaking, it would appear that they are good enough for most practical purposes when N_{SO} is greater than about five. The actual value of N_{SO} is always less than that predicted by equation 4(31). which therefore gives conservative values of N_{SO} and Q_O.

In design we know the overall separation Q_O and overall operating conditions J_1J_2 $(= m_1m_2G/L)$ and so N_{SO} may be obtained from Fig. 4.4 and J_2 from Fig. 4.5. The value of N_{SO} corresponds to the minimum total number of stages and $J_2 = m_2G/S$ gives the optimum flowrate of the intermediate solvent S. In operation we know N_{SO} (if $N_{S1} = N_{S2}$) and J_1J_2 and so the maximum separation Q_O follows from Fig. 4.4 and J_2 (and hence S) from Fig. 4.5.

When $J_1J_2 = 1$ writing $J_2 = 1/J_1$ shows that the governing equations are only satisfied by $J_1 = J_2 = 1$, $N_{S1} = N_{S2} = N_{SO}$ and $Q_1 = Q_2 = Q_O = 1$. However, if equation 4(10) is re-expressed in terms of $Q_{2/4}$ (defined by equation 4(29)) as in Appendix 4b, substituting $J_1 = J_2 = 1$ yields,

$$N_{SO} = 2\,(Q_{2/4} - 1) \qquad 4(32)$$

which is an exact result analogous to that for the differential case given in equation 4(30). Other limiting cases are discussed in Appendix 4i.

Q_O may be re-expressed in terms of any of the other separation factors listed in Table 2.1 using the overall balance involving J_1J_2. For example, equation 4(31) (which applies except when the value of N_{SO} is small) may be rewritten:

$$N_{SO} \ln 1/\sqrt{(J_1J_2)} = \ln\,[(1 - J_1J_2)\,Q_{2/4} + J_1J_2] \qquad 4(33)$$

which reduces to equation 4(32) when $J_1J_2 = 1$. Equation 4(33) is analogous to that derived by Kremser(48) for a single contactor.

4.4. Comparison with Ordinary Counter-current Extraction

It is interesting to compare forward and back extraction using counter current flow with ordinary counter-current extraction using no intermediate solvent.

Differential Case

Counter-current extraction is governed by equations 2(11) and (12) which may be written

$$N_T(1 - J) = \ln Q \qquad 4(34)$$

where $N_T = N_G = N_L/J$ and $J = mG/L$ and Q is defined in identical fashion to equation 4(26). Optimum conditions in counter-current forward and back extraction are governed by equation 4(25) which is

$$N_{TO}(1 - \sqrt{(J_1 J_2)}) = \ln Q_O \qquad 4(35)$$

where $N_{TO} = N_{G2} = N_{L1}/J_1$ and $J_1 = J_2 = \sqrt{(J_1 J_2)}$ and $J_1 J_2 = m_1 m_2\, G/L$ and Q_O is defined by equation 4(26).

Comparing the two types of extraction under the same conditions of individual extraction such that $J = J_1 = J_2$ shows that when the separations Q and Q_O are equal, so are the number of transfer units N_T and N_{TO}. However, it must be remembered that this comparison is not favourable to counter-current extraction as there are N_{TO} transfer units in both the forward and back extractors.

Alternatively the two types of extraction could be compared under the same conditions of overall extraction such that $m = m_1 m_2$ or $J = J_1 J_2$. Then for a given separation $Q = Q_O$

$$\frac{N_T}{N_{TO}} = \frac{1}{1 + \sqrt{(J_1 J_2)}} \qquad 4(36)$$

and for a given number of transfer units $N_T = N_{TO}$,

$$\frac{\ln Q_O}{\ln Q} = \frac{1}{1 + \sqrt{(J_1 J_2)}} \qquad 4(37)$$

These ratios are always less than unity, showing that counter-current extraction is always more efficient than forward and back extraction for the differential case.

Stagewise Case

Counter-current extraction in a contactor having N_S stages is governed by equations 2(11) and (12) which may be written

$$N_S \ln 1/J = \ln Q \qquad 4(38)$$

where $J = mG/L$ and Q is defined in identical fashion to equation 4(26). For most practical purposes optimum conditions in counter-current forward and back extraction are governed by

$$N_{SO} \ln 1/\sqrt{(J_1 J_2)} = \ln Q_O \tag{4(39)}$$

where $N_{SO} = N_{S1} = N_{S2}$ and $J_1 J_2 = m_1 m_2 G/L$ and Q_O is defined by equation 4(26).

The two types of extraction may be compared under the same conditions of overall extraction $J = J_1 J_2$. Thus when $J = J_1 J_2$ and $Q = Q_O$,

$$\frac{N_S}{N_{SO}} = \frac{1}{2} \tag{4(40)}$$

and similarly when $J = J_1 J_2$ and $N_S = N_{SO}$

$$\frac{\ln Q_O}{\ln Q} = \frac{1}{2} \tag{4(41)}$$

When N_{SO} is small the ratios N_S/N_{SO} and $\ln Q_O/\ln Q$ lie between 0·5 and 1·0 showing that ordinary counter-current extraction is always more efficient than forward and back counter-current extraction.

When N_{SO} is large it is also possible to compare the two types of extraction under the same condition of individual extraction $J = J_1 = J_2$. This shows that the separations Q and Q_O are equal if the number of stages N_S and N_{SO} are equal.

Example 4.1

Uranium is leached from ore with H_2SO_4 to a concentration of 1 g U_3O_8/l. as discussed by Grinstead, Shaw and Long(15). Ninety-nine per cent of the uranium is to be extracted from the H_2SO_4 leach liquors with 0·1 M dodecyl phosphoric acid (DDPA) in kerosene and the organic phase is to be stripped with 1 l. of uranium-free 10 M HCl/50 l. of leach liquor. Assuming the equilibrium data are such that for forward extraction $m_1 = 10$ and $c_1 = 0$, and for back extraction $m_2 = 30$ and $c_2 = 0$, what is the minimum number of stages required and the optimum recirculation rate of the DDPA?

SOLUTION

In this example $x_{in} = 1$ g/l, $x_{out} = 0{\cdot}01$ g/l, $y_{in} = 0$ g/l and $G/L = 1/50$. From an overall balance $y_{out} = 50 \times 0{\cdot}99 = 49{\cdot}5$ g/l. The values of J_1J_2 and Q_O are obtained below,

$$J_1J_2 = m_1m_2G/L = 10 \times 30 \times 1/50 = 6$$

so that

$$(J_1J_2)^* = 1/J_1J_2 = 0{\cdot}17,$$

and

$$Q_O = \frac{y_{in} - y(x_{out})}{y_{out} - y(x_{in})} = \frac{y_{in} - m_1m_2x_{out}}{y_{out} - m_1m_2x_{in}} = \frac{0 - 3}{49{\cdot}5 - 300} = 0{\cdot}01195$$

so that

$$Q_O^* = 1/Q_O = 83{\cdot}6.$$

Using Figs. 4.4 and 4.5, $N_{SO} = 4{\cdot}8$ and $J_2^* = 1/J_1 = 0{\cdot}35$ (equation 4(24) predicts $N_{SO} = 4{\cdot}93$). Thus $J_1 = 2{\cdot}86$ and $S/L = J_1/m_1 = 2{\cdot}86/10 = 0{\cdot}286$.

For contactor 1 using equation 4(2) with $N_1 = 4{\cdot}8$ and $J_1 = 2{\cdot}86$ gives $1/Q_1 = 154 = \dfrac{z_t - m_1x_{in}}{z_b - m_1x_{out}} = \dfrac{z_t - 10}{z_b - 0{\cdot}1}$. An overall balance round contactor 1 gives $z_t - z_b = 0{\cdot}99/0{\cdot}286 = 3{\cdot}46$. Thus $z_b = 0{\cdot}058$ g U_3O_8/l and $z_t = 3{\cdot}52$ g U_3O_8/l.

For the operation of a forward and back extraction system with $N_1 = N_2 = 4{\cdot}8$ and $J_1J_2 = 6$ reversing the calculation gives the maximum separation $1/Q_O = 83{\cdot}6$ and the optimum value of $J_1 = 2{\cdot}86$.

For a differential process the optimum recirculation rate is given by $J_1 = J_2 = \sqrt{(J_1J_2)} = 2{\cdot}45$ and so the minimum number of transfer units when $1/Q_O = 83{\cdot}6$ is given by Fig. 4.3 or equation 4(19) as $N_{L1} = 7{\cdot}5$ and so $N_{G2} = N_{L1}/J_1 = 3{\cdot}06$.

Under the same conditions of overall extraction (so that $J = J_1J_2$) the number of ordinary counter-current stages required to give the same separation in a single contactor is given by equation 4(38) as $N_S = \ln Q_O/\ln 1/J_1J_2 = 2{\cdot}48$ so that $N_S/N_{SO} = 0{\cdot}516$. The number of transfer units is given by equation 4(34) as $N_T = N_G = N_L/J_1J_2 = \ln Q_O/(1 - J_1J_2) = 0{\cdot}886$ so that $N_T/N_{TO} = 0{\cdot}29$ as indicated by equation 4(36).

Example 4.2

In the reprocessing of irradiated fuel, uranium is to be extracted from 3N HNO_3 into 20% tributyl phosphate in odourless kerosene (TBP/OK) and then backwashed into 0·01N HNO_3 (initially not containing uranium), using pulse columns as described by Logsdail and Larner[9]. The feed solution contains 300 g U/l and this must be reduced to 1 mg U/l when the flow of 3N acid is 3·33 m³/day and of 0·01N acid 16·7 m³/day. Assuming the equilibrium data are such that for forward extraction $m_1 = 20$ and $c_1 = 0$, and for back extraction $m_2 = 100$ and $c_2 = 0$, what is the minimum total number of transfer units and optimum recirculation rate of the TBP/OK?

SOLUTION

In this example $x_{in} = 300$ g U/l, $x_{out} = 0{\cdot}001$ g U/l, $y_{in} = 0$ g U/l and $G/L = 16{\cdot}7/3{\cdot}33$ so from an overall balance $y_{out} = 60$ g U/l.

The overall extraction factor is given by $J_1J_2 = m_1m_2G/L = 20 \times 100 \times 16{\cdot}7/3{\cdot}33 = 10^4$ so the optimum values of $J_1 = J_2 = J_O = \sqrt{(J_1J_2)} = 100$. Thus $S/L = J_1/m_1 = 100/20 = 5$ and the optimum recirculation rate of intermediate solvent is $S = 5 \times 3{\cdot}33 = 16{\cdot}7$ m³/day.

The overall separation factor is given by

$$Q_O = \frac{y_{in} - m_1m_2x_{out}}{y_{out} - m_1m_2x_{in}} = \frac{0 - 2000 \times 0{\cdot}001}{60 - 2000 \times 300} = \frac{1}{3 \times 10^5}$$

and so $Q_O^* = 1/Q_O = 3 \times 10^5$.

The optimum value of $J_O^* = 1/J_O = 0{\cdot}01$ and the minimum number of transfer units are given by

$$N_{L1} = \ln Q_O^*/(1 - J_O^*) = \ln(3 \times 10^5)/0{\cdot}99 = 12{\cdot}7$$

and

$$N_{G2} = N_{L1}/J_1 = 12{\cdot}7/100 = 0{\cdot}127$$

As $Q_1 = Q_O$ the separation in contactor 1 is given by

$$\frac{1}{Q_1} = 3 \times 10^5 = \frac{z_t - m_1x_{in}}{z_b - m_1x_{out}} = \frac{z_t - 6000}{z_b - 0{\cdot}020}$$

An overall balance round contactor 1 gives

$$z_t - z_b = (x_{in} - x_{out})\, L/G = 300/5 = 60$$

and so

$$z_t = 60 \text{ g U/l} \quad \text{and} \quad z_b = 0{\cdot}00002 \text{ g U/l}$$

In the operation of this forward and back extraction system with $N_{L1} = 12{\cdot}7$ and $N_{G2} = 0{\cdot}127$ reversing the calculation gives the optimum recirculation rate $S = 16{\cdot}7$ m^3/day and the maximum separation $Q_O = 3 \times 10^5$.

Under the same conditions of overall extraction (so that $J = J_1 J_2$) the number of transfer units required to give the same separation $Q = Q_O$ is given by equation 4(34) as $N_T = N_G = N_L/J_1J_2 = \ln Q_O/(1 - J_1J_2) = 0{\cdot}00126$ and so $N_T/N_{TO} = 0{\cdot}0099$ as indicated by equation 4(36).

4.5. Operation of an Existing Contactor

In an existing contactor it is not possible to optimise the operation with respect to N_1 and N_2 as their values are fixed. The value of $(\partial Q_O/\partial N_1)_{J_1}$ is not equal to zero and so equations 4(15) and (16) do not apply. However, it is still possible to find the optimum recirculation rate S (i.e. J_1 or J_2) which gives the maximum value of Q_O for the given values of N_1 and N_2 and given operating conditions L, G, m_1 and m_2 (i.e. J_1J_2). Setting $(\partial Q_O/\partial J_1)_{N_1}$ equal to zero, equations 4(17) and (18) still hold and there are now five equations—4(2), (6), (10), (12) and (17) or (18)—and five unknowns—J_1, J_2, Q_1, Q_2 and Q_O if the values of N_1, N_2 and J_1J_2 are fixed.

Differential Case

Substitution of equations 4(2) and (6) shows that equation 4(18) may be rewritten

$$\frac{Q_1 - 1 - Q_1 \ln Q_1}{(Q_1 - 1)^2} = \frac{Q_2 - 1 - Q_2 \ln Q_2}{(Q_2 - 1)^2} \qquad 4(42)$$

so that Q_1 equals Q_2 and it follows from equation 4(10) that both must equal Q_O. Equations 4(2) and (6) may thus be rewritten

$$\frac{N_1}{J_1}(1 - J_1) = \ln Q_O \tag{4(43)}$$

and

$$N_2(1 - J_2) = \ln Q_O \tag{4(44)}$$

which may be rearranged in terms of J_1J_2 rather than J_1 and J_2 to give

$$\left(\frac{J_1J_2}{N_1} + \frac{1}{N_2}\right)\ln Q_O = 1 - J_1J_2 \tag{4(45)}$$

and

$$\left(1 - \frac{\ln Q_O}{N_1/J_1}\right)\left(1 - \frac{\ln Q_O}{N_2}\right) = J_1J_2 \tag{4(46)}$$

These equations enable the maximum overall separation Q_O to be obtained for an existing contactor with fixed N_1 and N_2 and given overall operating conditions J_1J_2. Knowledge of Q_O gives J_1 and J_2 from which S, the recirculation rate of the intermediate solvent may be calculated.

Equations 4(45) and (46) are plotted in Appendix 4j for values of $J_1J_2 = 0{\cdot}1$, $0{\cdot}3$, $0{\cdot}5$, $0{\cdot}7$ and $0{\cdot}9$, These figures show that plots of N_1/J_1 against N_2 for different values of Q_O at constant J_1J_2 are symmetrical and plots of N_1 against N_2 become more symmetrical as J_1J_2 approaches one.

When N_1, N_1/J_1 or N_2 approach infinity at finite Q_O, J_1 or J_2 must approach unity. In the symmetrical curves when N_1/J_1 or N_2 approach infinity, N_2 or N_1/J_1 respectively both approach $\ln Q_O/(1 - J_1J_2)$. This same limit is approached by N_2 when N_1 tends to infinity in the unsymmetrical curves but when N_2 tends to infinity N_1 approaches $J_1J_2 \ln Q_O/(1 - J_1J_2)$. When N_1/J_1 equals N_2 equation 4(46) reduces to equation 4(25).

Stagewise Case

Inspection of equations 4(2), (6) and (17) does not reveal any simple solution as for the differential case. The governing equations were therefore solved as shown in Appendix 4k using the digital computer programs described in Appendices 4o and 4p. Because we require to find the values of Q_O and J_1 from given values of N_1, N_2 and J_1J_2 the solution is difficult to present concisely. Plots of N_1 and N_1/J_1 against

N_2 for different values of Q_O at constant J_1J_2 are given in Appendix 41. In contrast to the differential case discussed above curves of N_1 against N_2 are symmetrical and curves of N_1/J_1 against N_2 become more symmetrical as the value of J_1J_2 approaches unity. Log Q_O is used as the constant parameter as this gives an even gradation.

Knowing the numbers of stages N_1 and N_2 and overall operating conditions J_1J_2 ($= m_1m_2G/L$) gives the maximum overall separation Q_O by interpolation between the sets of curves of N_1 versus N_2. This gives N_1/J_1 from the curves of N_1/J_1 versus N_2, which enables J_1 ($= m_1S/L$) to be determined and hence the optimum flowrate of the intermediate solvent S.

As N_1, N_1/J_1 or N_2 approach infinity the results of the computer program show that J_1 or J_2 approach unity so that Q_1 and Q_2 remain finite as for the differential case. When $J_1 = 1$ so $J_2 = J_1J_2$ and it follows from equation 4(10) that $Q_2 = Q_O$; when $J_2 = 1$ it follows from equation 4(10) that $Q_1 = Q_O$. Equation 4(6) thus indicates that as N_1 or N_1/J_1 approaches infinity, so N_2 approaches $\ln Q_O/\ln 1/J_1J_2$. Similarly equation 4(2) indicates that this limit is also approached by N_1 as N_2 approaches infinity but that N_1/J_1 approaches $\ln Q_O/(J_1J_2 \ln 1/J_1J_2)$.

These limits and the differential result of equations 4(45) and (46) suggest the approximations,

$$\left(\frac{1}{N_1} + \frac{1}{N_2}\right) \ln Q_O = \ln 1/J_1J_2 \qquad 4(47)$$

and

$$\left(1 - \frac{(J_1J_2)^{\mu-1} \ln Q_O}{N_1/J_1}\right)\left(1 - \frac{(J_1J_2)^{\mu} \ln Q_O}{N_2}\right) = 1 - (J_1J_2)^{\mu} \ln 1/J_1J_2 \qquad 4(48)$$

where the index μ can have any value. Furthermore, equation 4(47) reduces to equation 4(31) when $N_1 = N_2$ and equation 4(48) satisfies the exact limit that $Q_O = 1$ when $J_1J_2 = 1$.

Equations 4(47) and (48) closely represent the exact solutions plotted in Appendix 41 when

$$\mu = 0{\cdot}75 \left(\frac{\ln Q_O}{N_2 \ln 1/J_1J_2}\right)^2$$

so that μ approaches 0·75 when N_1/J_1 is large and zero when N_2 is large. This variation reproduces the two-valued nature of the curves of N_1/J_1 versus N_2 at low values of J_1J_2. Even better agreement may be obtained by using more complicated expressions for μ. As equation 4(48) is empirical it cannot be inverted.

The approximations become more exact as J_1J_2 approaches unity and Q_O increases. In general the approximate curves tend to lie above the corresponding exact curves for the same value of Q_O. This can give rise to large errors in the values of J_1, estimated from equation 4(48), when the curves of N_1/J_1 versus N_2 become horizontal. However, values of Q_O estimated from equation 4(47) are usually very close to the exact value. Thus in general the approximations predict the maximum possible separation more accurately than the optimum recirculation rate needed to achieve it.

Example 4.3

Gasolene is to be sweetened with an intermediate solvent containing 30% NaOH and 20% organic acids, at 90°F in a plant containing the equivalent of two theoretical extraction stages. The stripper contains the equivalent of three theoretical stages and uses 10 lb of steam per 42 United States gallons of treated gasolene.

The mercaptan contents of typical straightrun and cracked gasolene and the corresponding values of m_1 and m_2 (expressed as a ratio of weight volumes) are given by Happel, Cauley and Kelly[16] as indicated in Table 4.2. What are the optimum operating conditions?

TABLE 4.2. MERCAPTAN CONTENTS AND EQUILIBRIUM DATA FOR STRAIGHT-RUN AND CRACKED GASOLENES

Mercaptan	Fraction of total mercaptan sulphur		m_1	m_2
	Straightrun	Cracked		
Methyl	0·04	0·19	2380	0·94
Ethyl	0·06	0·34	570	1·95
Propyl	0·13	0·18	170	3·25
Butyl	0·19	0·15	77	3·60
Amyl	0·18	0·09	51	2·75
Hexyl	0·40	0·05	36	1·40
Total mercaptan sulphur % by weight	0·0265	0·0357		

SOLUTION

The optimum recirculation rate, S, will be determined for the hexyl mercaptan which has the lowest $m_1 m_2$ product (and so is the most difficult to extract). Thus $J_1 J_2 = m_1 m_2 G/L = 36 \times 1{\cdot}40 \times 1/35 = 1{\cdot}44$ (as $G/L = 1$ Imperial gallon condensate/35 Imperial gallons of gasolene).

Solution of equations 4(2), (6), (12) and (17) with $N_{S1} = 2$, $N_{S2} = 3$ and $J_1 J_2 = 1{\cdot}44$ gives $J_2 = 0{\cdot}639$ and $J_1 = 2{\cdot}25$. Thus $Q_1 = 1/J_1^{N_1} = 0{\cdot}198$ and $Q_2 = 1/J_2^{N_2} = 3{\cdot}83$ and using equation 4(10) gives $Q_O = 0{\cdot}61$. The approximate equations 4(47) and (48) predict $Q_O = 0{\cdot}645$ and $J_1 = 1{\cdot}26$ illustrating the fact that the estimate of Q_O is reasonably accurate even when that of J_1 is not so accurate. This example is a severe test of the approximations as the value of Q_O is so low.

As $y_{\text{in}} = 0$, $Q_O = x_{\text{out}}/x(y_{\text{out}})$ and from an overall balance $J_1 J_2 x(y_{\text{out}}) = x_{\text{in}} - x_{\text{out}}$. Eliminating $x(y_{\text{out}})$ gives $x_{\text{out}}/x_{\text{in}} = (J_1 J_2 - 1) Q_O / (J_1 J_2 - Q_O)$ and so for the hexyl mercaptan $x_{\text{out}}/x_{\text{in}} = (1{\cdot}44 - 1) \times 0{\cdot}61/(1{\cdot}44 - 0{\cdot}61) = 0{\cdot}322$.

Now $S/L = J_1/m_1 = 2{\cdot}25/36 = 0{\cdot}0625$ and $G/S = J_2/m_2 = 0{\cdot}639/1{\cdot}40 = 0{\cdot}455$, so J_1 and J_2 may be obtained for the other mercaptans. Q_1 and Q_2 follow from equations 4(2) and (6), and hence Q_O from equation 4(10). The resulting values of $x_{\text{out}}/x_{\text{in}}$ are given in Table 4.3.

Typical operating conditions[16] are $S/L = 0{\cdot}15$ and corresponding values of $x_{\text{out}}/x_{\text{in}}$ may be obtained as described immediately above. These are quoted by Treybal[3] and are also listed in Table 4.3.

It can be seen that for the optimum operating conditions the ratio $x_{\text{out}}/x_{\text{in}}$ is significantly lower for each mercaptan than with the typical operating conditions. The fraction of total mercaptan sulphur remaining is reduced from 0·183 to 0·163 for the straightrun gasolene and from 0·054 to 0·044 for the cracked gasolene. The values of z_b and z_t for each mercaptan may be obtained from Q_1 or Q_2 and an overall balance round each contactor as indicated in Example 4.1. For the hexyl mercaptan in the gasolene the optimum S/L of 0·0625 gives $z_b = 0{\cdot}087$ and $z_t = 0{\cdot}202$, whereas the typical S/L of 0·15 gives $z_b = 0{\cdot}127$ and $z_t = 0{\cdot}172$.

Reducing the typical recirculation rate S by a factor of 2·4 will reduce the operating costs and, in addition, give a better extraction of the mercaptans. However, a solvent ratio of 1/16 may present distribution problems and this could be the reason for the use of a higher solvent flow.

TABLE 4.3. FRACTION OF MERCAPTAN SULPHUR REMAINING IN STRAIGHTRUN AND CRACKED GASOLENES UNDER OPTIMUM AND TYPICAL OPERATING CONDITIONS

Mercaptan	Optimum $S/L = 0{\cdot}0625$			Typical $S/L = 0{\cdot}15$		
	$\frac{x_{out}}{x_{in}}$	Fraction of mercaptan sulphur remaining		$\frac{x_{out}}{x_{in}}$	Fraction of mercaptan sulphur remaining	
		Straightrun	Cracked		Straightrun	Cracked
Methyl	0·0093	0·000372	0·00177	0·0127	0·000509	0·00241
Ethyl	0·0125	0·000750	0·00425	0·0205	0·00123	0·00697
Propyl	0·0187	0·00243	0·00337	0·0319	0·00415	0·00574
Butyl	0·0561	0·0107	0·00844	0·0611	0·0116	0·00918
Amyl	0·112	0·0202	0·0101	0·1332	0·0240	0·0120
Hexyl	0·322	0·129	0·0161	0·354	0·1415	0·0177
Total		0·163	0·044		0·183	0·054
Total mercaptan sulphur remaining (% by weight)		0·00432	0·00157		0·00485	0·00193

Example 4.4

An acid gas containing 0·014 weight ratio CO_2 is to be purified with an aqueous ethanolamine solution in an absorber containing the equivalent of 2·5 theoretical stages as described by Kohl and Reisenfeld.[(2)] The stripping column contains the equivalent of eight theoretical stages and uses 1 lb of steam/20 lb of CO_2 free acid gas. Assuming that the equilibrium relationships (with concentrations in weight ratios) may be represented by $z = 1{\cdot}18x + 0{\cdot}04$ and $y = 5{\cdot}1z$, for the absorber and stripper respectively, calculate the maximum separation and optimum rate of recirculation of the ethanolamine solution.

SOLUTION

In this example $N_{S1} = 2{\cdot}5$, $N_{S2} = 8$, $y_{in} = 0$, $x_{in} = 0{\cdot}014$ w/w, $G/L = 1/20$, $m_1 = 1{\cdot}18$ and $m_2 = 5{\cdot}1$.

The overall extraction factor is given by

$$J_1 J_2 = m_1 m_2 G/L = 1{\cdot}18 \times 5{\cdot}1/20 = 0{\cdot}3.$$

Solution of equations 4(2), (6), (12) and (17) thus gives $J_1 = 0{\cdot}4725$, $J_2 = 0{\cdot}635$, $Q_1 = 6{\cdot}56$ and $Q_2 = 38{\cdot}0$. Substitution of these values in equation 4(10) gives $Q_O = 11{\cdot}0$ and the flow ratios are $S/L = J_1/m_1 = 0{\cdot}4725/1{\cdot}18 = 0{\cdot}400$ and $G/S = J_2/m_2 = 0{\cdot}635/5{\cdot}1 = 0{\cdot}1245$.

Figure A41.3 with $N_1 = 2{\cdot}5$ and $N_2 = 8$ indicates that $\log_{10} Q_O \approx 1{\cdot}0$ (that is $Q_O \approx 10$) and so from Fig. A41.4, $N_1/J_1 \approx 5$ giving $J_1 \approx 2{\cdot}5/5 = 0{\cdot}5$. The approximate equations 4(47) and (48) predict $Q_O \approx 10$ and $J_1 \approx 0{\cdot}41$.

The overall separation factor is defined by equation 4(9) as

$$Q_O = \frac{y_{in} - (m_1 m_2 + m_2 c_1 + c_2)}{y_{out} - (m_1 m_2 + m_2 c_1 + c_2)}$$

so

$$11{\cdot}0 = \frac{0 - (1{\cdot}18 \times 5{\cdot}1 x_{out} + 5{\cdot}1 \times 0{\cdot}04 + 0)}{y_{out} - (1{\cdot}18 \times 5{\cdot}1 \times 0{\cdot}014 + 5{\cdot}1 \times 0{\cdot}04 + 0)}$$

or

$$11 y_{out} + 6 x_{out} = 2{\cdot}97$$

An overall balance gives

$$20 x_{out} + y_{out} = 0{\cdot}28$$

and so

$$x_{out} = 0{\cdot}000514 \text{ lb } CO_2/\text{lb acid gas}$$

and

$$y_{out} = 0{\cdot}270 \text{ lb } CO_2/\text{lb steam}$$

Using equation 4(6) the individual separation factor is given by

$$Q_2 = \frac{y_{in} - m_2 z_b}{y_{out} - m_2 z_t}$$

so

$$38{\cdot}0 = \frac{0 - 5{\cdot}1 z_b}{0{\cdot}270 - 5{\cdot}1 z_t}$$

or

$$10{\cdot}25 = 194 z_t - 5{\cdot}1 z_b$$

A balance round contactor 2 gives

$$z_t - z_b = (y_{out} - y_{in})\, G/S = (0{\cdot}270 - 0) \times 0{\cdot}1245 = 0{\cdot}0336$$

and so $z_t = 0{\cdot}0534$ and $z_b = 0{\cdot}0197$ lb CO_2/lb ethanolamine solution.

To summarise, the optimum recirculation rate of ethanolamine solution is 0·400 lb/lb of CO_2 free acid gas and the optimum concentrations are (in weight ratios) $x_{out} = 0{\cdot}000514$, $y_{out} = 0{\cdot}270$, $z_t = 0{\cdot}0534$ and $z_b = 0{\cdot}0197$.

For a differential process with $N_{L1} = 2{\cdot}5$, $N_{G2} = 8$ and $J_1 J_2 = 0{\cdot}3$, equations 4(45) and (46) give $Q_O = 17{\cdot}4$ and $J_1 = 0{\cdot}47$ or $S/L = 0{\cdot}40$ lb ethanolamine/lb CO_2 free acid gas.

Notation for Chapter 4

c	intercept of equilibrium line on y-axis
G	flowrate of phase in which solute concentration is y
J_1	extraction factor in contactor 1 ($J_1 = m_1 S/L$)
J_2	extraction factor in contactor 2 ($J_2 = m_2 G/S$)
K	overall extraction factor ($K = J_1 J_2 = m_1 m_2 G/L$)
L	flowrate of phase in which solute concentration is x
m_1	slope of equilibrium line in contactor 1 (equation $z = m_1 x + c_1$)
m_2	slope of equilibrium line in contactor 2 (equation $y = m_2 z + c_2$)

N number of theoretical stages or transfer units
N_G number of overall transfer units based on phase of flowrate G
N_L number of overall transfer units based on phase of flowrate L
N_S number of theoretical stages
N_T number of transfer units
P total number of stages or transfer units in contactors 1 and 2
Q_o overall separation factor defined by equations 4(9) and (26) $(Q_o = (y_{in} - y(x_{out}))/(y_{out} - y(x_{in})))$
Q_O optimum value of overall separation factor
$Q_{2/4}$ separation factor defined by equation 4(29) $(Q_{2/4} = (y_{in} - y(x_{in}))/(y_{out} - y(x_{in})))$
S flowrate of intermediate solvent (in which solute concentration is z)
x solute concentration in phase of flowrate L
$x(y)$ x concentration in equilibrium with concentration y $(x(y) = y/m_1m_2 - c_1/m_1 - c_2/m_1m_2)$
$x(z)$ x concentration in equilibrium with concentration z $(x(z) = z/m_1 - c_1/m_1)$
y solute concentration in phase of flowrate G
$y(x)$ y concentration in equilibrium with concentration x $(y(x) = m_1m_2x + m_2c_1 + c_2)$
$y(z)$ y concentration in equilibrium with concentration z $(y(z) = m_2z + c_2)$
z solute concentration in phase of flowrate S
$z(x)$ z concentration in equilibrium with concentration x $(z(x) = m_1x + c_1)$
$z(y)$ z concentration in equilibrium with concentration y $(z(y) = y/m_2 - c_2/m_2)$

Greek symbol

μ index of J_1J_2 (in approximate solution for existing stagewise contactor)

Subscripts

in refers to concentration before inlet of contactor
out refers to concentration after outlet of contactor
b refers to intermediate solvent leaving contactor 2

G	refers to phase of flowrate G
L	refers to phase of flowrate L
o	refers to overall separation
O	refers to optimum conditions
t	refers to intermediate solvent leaving contactor 1
1	refers to contactor 1
2	refers to contactor 2

Superscript

*	denotes inversion.

The above quantities may be expressed in any set of consistent units in which force and mass are not defined independently.

References

1. L. Alders, *Liquid–Liquid Extraction*, 2nd ed., p. 76, Elsevier Publishing Company, Amsterdam, 1959.
2. A.L. Kohl and F.C. Reisenfeld, *Gas Purification*, McGraw-Hill Book Company Inc., New York, 1960.
3. R.E. Treybal, *Liquid Extraction*, 1st ed., Chap. 8, McGraw-Hill Book Co. Inc., New York, 1951.
4. R.F. Ruthruth and D.F. Wilcock, Solvent extraction of vegetable drying oils, *Trans. Am. Inst. Chem. Engrs.* **37,** 649 (1941).
5. S.W. Gloyer, Furans in vegetable oil refining, *Ind. Eng. Chem.* **40,** 228 (1948).
6. F.R. Bruce *et al.* (Editors), *Progress in Nuclear Energy*, Series III. *Process Chemistry*, Vol. 1, Chap. 5, Wet Processes for Radiochemical Separations, McGraw-Hill Book Company Inc., New York, 1956.
7. D.G. Kirraker, Temperature effects on TBP solvent extraction processes, *Proceedings of Second United Nations International Conference on the Peaceful Uses of Atomic Energy*, Vol. 17, p. 333, United Nations, Geneva, 1958.
8. J.R. Flanary, Solvent extraction separation of uranium and plutonium from fission products by means of tributyl phosphate, *Proceedings of the International Conference on the Peaceful Uses of Atomic Energy*, Vol. 9, p. 528, United Nations, New York, 1956.
9. D.H. Logsdail and G.S. Larner, *Processing in Limited Geometry.* Part VII. High Throughputs in Pulsed Plate Columns for Uranium, Tributyl Phosphate Systems. A.E.R.E.-R4408 (unclassified).
10. K.B. Brown, C.F. Coleman, D.J. Grouse, C.A. Blake and A.D. Ryan, Solvent extraction processing of uranium and thorium ores, *Proceedings of Second United Nations International Conference on the Peaceful Uses of Atomic Energy*, Vol. 3, p. 472, United Nations, Geneva, 1958.
11. S. Glasstone, *Principles of Nuclear Reactor Engineering*, pp. 416 and 448, D. Van Nostrand & Co. Ltd., London, 1955.
12. Albany, *Chemical Engineering* **62** (10), 112 (1955).

13. R. S. Long, D. A. Ellis and R. H. Bailes, Recovery of uranium from phosphates by solvent extraction, *Proceedings of the International Conference of the Peaceful Uses of Atomic Energy*, Vol. 8, p. 77, United Nations, New York, 1956.
14. M. Benedict and T. H. Pigford, *Nuclear Chemical Engineering*, McGraw-Hill Book Company Inc., New York, London, 1957.
15. R. K. Grinstead, K. W. Shaw and R. S. Long, Solvent extraction of uranium from acid leach slurries and solutions, *Proceedings of the International Conference of the Peaceful Uses of Atomic Energy*, Geneva, 1955, Vol. 8, p. 71, United Nations, New York, 1956.
16. J. Happel, S. P. Cauley and H. S. Kelly, Critical analysis of sweetening processes and mercaptan removal, *Oil and Gas Journal* **41** (27), 136 (1942).
17. D. L. Yabroff and E. R. White, Action of solutizers in mercaptan extraction, *Ind. Eng. Chem.* **32,** 950 (1940).
18. J. M. Page, C. C. Buchler and S. H. Diggs, Production of lubricating oils by extraction with dichloroethyl ether, *Ind. Eng. Chem.* **25,** 418 (1933).
19. L. C. Stewart, Commercial extraction of bromine from sea water, *Ind. Eng. Chem.* **26,** 361 (1934).
20. R. E. Treybal, *Liquid Extraction*, 2nd ed., p. 551, McGraw-Hill Book Company Inc., New York, 1963.
21. T. K. Sherwood and R. L. Pigford, *Absorption and Extraction*, p.135, McGraw-Hill Book Company Inc., New York, 1951.
22. A. P. Colburn, *Collected Papers on the Teaching of Chemical Engineering, American Society for Engineering Education*, Summer School for Teaching of Chemical Engineering, Pennsylvania State College, 1936.
23. R. Aris, D. F. Rudd and R. R. Amundson, Optimum cross-current extraction, *Chem. Eng. Sci.* **12,** 88 (1960).
24. D. F. Rudd and E. D. Blum, Optimum cross-current extraction with product recycle, *Chem. Eng. Sci.* **17,** 277 (1962).
25. R. Aris, Optimum operating conditions in a series of stirred tank reactors, *Chem. Eng. Sci.* **13,** 75 (1960).
26. V. G. Jenson and G. V. Jeffries, Economic analysis of liquid/liquid extraction processes, *Brit. Chem. Eng.* **6** (10), 676 (1961).
27. J. M. Srygley and C. D. Holland, Optimum design of conventional and complex distillation columns. *A.I.Ch.E.J.* **11** (4), 695 (1965).
28. D. Luss, Optimum volume ratios for residence time in stirred tank reactor sequences, *Chem. Eng. Sci.* **20** (2), 171 (1965).
29. R. P. King, Calculation of optimal conditions for chemical reactors of combined type, *Chem. Eng. Sci.* **20** (6), 537 (1965)
30. R. Jackson, Variational solution of unsteady state optimisation in complex chemical plants, *Chem. Eng. Sci.* **20** (5), 405 (1965).
31. L. T. Fan, L. E. Erickson, R. W. Sucher and G. S. Matbas, Optimal design of sequences of continuous flow stirred-tank reactors with product recycle, *Ind. Eng. Chem. (Process Design and Development)* **4** (4), 431 (1965).
32. G. A. Almasy, J. J. Hay and I. M. Pallai, Ammonia synthesis loop optimisation, *Brit. Chem. Eng.* **11** (3), 188 (1966).
33. H. H.-Y. Chien, Optimisation of recycle problems, *Ind. Eng. Chem. (Fundamentals)* **5** (1), 66 (1966).
34. G. S. G. Beveridge and R. S. Schechter, Optimal reaction systems with recycle, *Ind. Eng. Chem. (Fundamentals)* 4 (3), 257 (1965).

35. C. W. Di Bella and W. F. Stevens, Process optimisation by non-linear programming, *Ind. Eng. Chem. (Process Design and Development)* **4** (1), 16 (1965).
36. M. W. McEwan and G. S. G. Beveridge, Process optimisation at the design stage with an objective function linear in profit and investment, *Chem. Eng. Sci.* **20** (12), 987 (1965).
37. M. M. Denn and R. Aris, Green's functions and optimal systems. Necessary conditions and iterative technique, *Ind. Eng. Chem. (Fundamentals)* **4** (19), 7 (1965).
38. I. Coward and R. Jackson, Optimum temperature profiles in tubular reactors: an exploration of some difficulties in the use of Pontryagin's Maximum Principle, *Chem. Eng. Sci.* **20** (10), 911 (1965).
39. C. J. Pings, Optimisation of initial compositions. Adiabatic equilibrium gas phase reactors in the presence of inerts. *Ind. Eng. Chem. (Fundamentals)* **4** (3), 260 (1965).
40. E. S. Lee, Quasi linearisation, non-linear boundary value problems and optimisation, *Chem. Eng. Sci.* **21** (2), 183 (1966).
41. M. M. Denn, R. D. Gray and J. R. Ferron, Optimisation in a class of distributed parameter systems, *Ind. Eng. Chem. (Fundamentals)* **5** (1), 59 (1966).
42. M. M. Denn and R. Aris, Green's functions and optimal systems. Complex interconnected structures, *Ind. Eng. Chem. (Fundamentals)* **4** (3), 248 (1965).
43. D. L. Yabroff, E. R. White and A. V. Caselli, The regeneration step in the solutizer process for the extraction of mercaptans, *Proc. A.P.I.* **20** (3), 57 (1939).
44. H. S. Mickley, T. K. Sherwood and C. E. Reed, *Applied Mathematics in Chemical Engineering*, McGraw-Hill Book Co. Inc., New York, 1957.
45. J. Happel, *Chemical Process Economics*, Wiley, New York, 1958.
46. A. P. Colburn, The simplified calculation of diffusional processes. General consideration of two film resistances, *Trans. Am. Inst. Chem. Engrs.* **35**, 211 (1939).
47. J. H. Perry, C. H. Chilton and S. D. Kirkpatrick (Editors), *Chemical Engineers' Handbook*, 4th ed., McGraw-Hill Book Co. Inc., New York, 1952.
48. A. Kremser, Theoretical analysis of absorption process, *Nat. Pet. News* **22**, 42 (1930).
49. G. C. Coggan, Multifit—a complete procedure for fitting non-linear static mathematical models to data, *Proceedings of Symposium on Efficient Computer Methods for the Practising Chemical Engineer*, Institution of Chemical Engineers Symposium Series No. 23, p. 32, London, 1967.

APPENDIX 4a
THE RELATIONSHIP BETWEEN OVERALL AND INDIVIDUAL SEPARATION FACTORS IN FORWARD AND BACK EXTRACTION

The overall separation factor Q_o is defined by equation 4(9) as

$$Q_o = \frac{y_{\text{in}} - (m_1 m_2 x_{\text{out}} + m_2 c_1 + c_2)}{y_{\text{out}} - (m_1 m_2 x_{\text{in}} + m_2 c_1 + c_2)} \qquad \text{A4a(1)}$$

so

$$Q_o - 1 = \frac{y_{in} - y_{out} + m_1 m_2 x_{in} - m_1 m_2 x_{out}}{y_{out} - (m_1 m_2 x_{in} + m_2 c_1 + c_2)} \qquad \text{A4a(2)}$$

Using the overall balance round contactors 1 and 2

$$x_{out} - x_{in} = (G/L)(y_{in} - y_{out}) \qquad \text{A4a(3)}$$

and the definition $J_1 J_2 = m_1 m_2 G/L$ this becomes

$$Q_o - 1 = \frac{(y_{in} - y_{out})(1 - J_1 J_2)}{y_{out} - (m_1 m_2 x_{in} + m_2 c_1 + c_2)} \qquad \text{A4a(4)}$$

or

$$\frac{1 - J_1 J_2}{Q_o - 1} = \frac{y_{out} - (m_1 m_2 x_{in} + m_2 c_1 + c_2)}{y_{in} - y_{out}} \qquad \text{A4a(5)}$$

The individual separation factor Q_1 defined by equation 4(4) is

$$Q_1 = \frac{z_b - (m_1 x_{out} + c_1)}{z_t - (m_1 x_{in} + c_1)} \qquad \text{A4a(6)}$$

so

$$Q_1 - 1 = \frac{z_b - z_t + m_1 x_{in} - m_1 x_{out}}{z_t - (m_1 x_{in} + c_1)} \qquad \text{A4a(7)}$$

Using the individual balance on contactor 1

$$z_b - z_t = (L/S)(x_{out} - x_{in}) \qquad \text{A4a(8)}$$

and the definition $J_1 = m_1 S/L$ this becomes

$$Q_1 - 1 = \frac{(x_{out} - x_{in})(1 - J_1)(L/S)}{z_t - (m_1 x_{in} + c_1)} \qquad \text{A4a(9)}$$

which combined with the overall balance A4a(3) and the definition $J_2 = m_2 G/S$ gives

$$Q_1 - 1 = \frac{(y_{in} - y_{out})(1 - J_1) J_2}{m_2 z_t - (m_1 m_2 x_{in} + m_2 c_1)} \qquad \text{A4a(10)}$$

or

$$\frac{(1 - J_1) J_2}{Q_1 - 1} = \frac{m_2 z_t - (m_1 m_2 x_{in} + m_2 c_1)}{y_{in} - y_{out}} \qquad \text{A4a(11)}$$

The individual separation factor Q_2 defined by equation 4(8) is

$$Q_2 = \frac{y_{\text{in}} - (m_2 z_b + c_2)}{y_{\text{out}} - (m_2 z_t + c_2)} \qquad \text{A4a(12)}$$

so

$$Q_2 - 1 = \frac{y_{\text{in}} - y_{\text{out}} + m_2 z_t - m_2 z_b}{y_{\text{out}} - (m_2 z_t + c_2)} \qquad \text{A4a(13)}$$

Using the individual balance on contactor 1

$$z_b - z_t = (G/S)(y_{\text{in}} - y_{\text{out}}) \qquad \text{A4a(14)}$$

and the definition $J_2 = m_2 G/S$ gives

$$Q_2 - 1 = \frac{(y_{\text{in}} - y_{\text{out}})(1 - J_2)}{y_{\text{out}} - (m_2 z_t + c_2)} \qquad \text{A4a(15)}$$

or

$$\frac{1 - J_2}{Q_2 - 1} = \frac{y_{\text{out}} - (m_2 z_t + c_2)}{y_{\text{in}} - y_{\text{out}}} \qquad \text{A4a(16)}$$

Combining equations A4a(5), (11) and (16) gives

$$\frac{1 - J_1 J_2}{Q_o - 1} = \frac{1 - J_2}{Q_2 - 1} + \frac{(1 - J_1) J_2}{Q_1 - 1} \qquad \text{A4a(17)}$$

which is equation 4(10).

APPENDIX 4b
COMPARISON WITH OTHER FORMULAE FOR FORWARD AND BACK EXTRACTION

Happel, Cauley and Kelly[18] derived for the stagewise case

$$\frac{x_{\text{out}}}{x_{\text{in}}} = \frac{(1 - J_2)(J_1^{N_1} - 1) + (J_2 - J_2^{N_2+1})(J_1 - 1)}{(1 - J_2)(J_1^{N_1} - 1) + (J_2 - J_2^{N_2+1})(J_1^{N_1+1} - 1)}$$

when c_1, c_2 and y_{in} are zero. Using equation 4(9) the overall separation factor becomes

$$Q_o = \frac{-m_1 m_2 x_{\text{out}}}{y_{\text{out}} - m_1 m_2 x_{\text{in}}}$$

and from an overall balance round contactors 1 and 2 of Fig. 4.1

$$y_{out} = (L/G)(x_{in} - x_{out})$$

so remembering that $J_1J_2 = m_1m_2G/L$ gives

$$Q_o = \frac{-J_1J_2x_{out}/x_{in}}{1 - J_1J_2 - x_{out}/x_{in}}$$

or

$$\frac{x_{out}}{x_{in}} = \frac{(1 - J_1J_2)Q_o}{Q_o - J_1J_2}$$

Substituting for Q_o from equation 4(10) leads to

$$\frac{x_{out}}{x_{in}} = \frac{(Q_2 - J_2)(Q_1 - 1) + (1 - J_1Q_1)(Q_2 - 1)J_2}{(Q_2 - J_2)(Q_1 - 1) + (1 - J_1)(Q_2 - 1)J_2}$$

which becomes the formula of Happel, Cauley and Kelly on writing $Q_1 = J_1^{-N_1}$ and $Q_2 = J_2^{-N_2}$ from equations 4(2) and (6) and rearranging.

APPENDIX 4c
USE OF OTHER SEPARATION FACTORS IN FORWARD AND BACK EXTRACTION

The separation factors Q_1, Q_2 and Q_o defined by equations 4(4), (8) and (9) respectively are of the form $Q_{3/4}$ as defined by the concentration differences in Table 2.1. Q_1 and Q_2 may be re-expressed in terms of any of the other separation factors through the formulae given in Table 2.3. For example, the relationship between $Q_{3/4}$ and $Q_{2/4}$ is given by formulae V and VII as

$$Q_{3/4} = (1 - J)Q_{2/4} + J \qquad \text{A4c(1)}$$

where J is J_1 or J_2.

Similarly Q_o may be re-expressed in terms of other overall separation factors. An overall balance round the two contactors in Fig. 4.1 gives

$$y(x_{in}) - y(x_{out}) = J_1J_2(y_{out} - y_{in}) \qquad \text{A4c(2)}$$

which shows that J_1J_2 replaces J in an overall balance round the single contactor in Fig. 1.1. The overall separation factor Q_o is defined in a closely similar manner to the separation factor $Q_{3/4}$ of Table 2.1 and may thus be expressed in terms of any of the other separation factors through the formulae given in Table 2.3 if J is replaced by $J_1 J_2$. For example,

$$Q_{3/4} = (1 - J_1 J_2)\, Q_{2/4} + J_1 J_2 \qquad \text{A4c(3)}$$

the overall separation factor $Q_{2/4}$ being defined by

$$Q_{2/4} = \frac{y_{\text{in}} - y(x_{\text{in}})}{y_{\text{out}} - y(x_{\text{in}})} \qquad \text{A4c(4)}$$

where

$$y(x_{\text{in}}) = m_1 m_2 x_{\text{in}} + m_2 c_1 + c_2$$

is the y concentration in equilibrium with z_t and this in turn is in equilibrium with x_{in}.

Re-expressing Q_1, Q_2 and Q_o in terms of $Q_{2/4}$ in equations 4(2), (6) and (10) gives

$$N_{S1} \ln 1/J_1 = \frac{N_{L1}}{J_1}(1 - J_1) = \ln\,[(1 - J_1)\, Q_1 + J_1] \qquad \text{A4c(5)}$$

$$N_{S2} \ln 1/J_2 = N_{G2}\,(1 - J_2) = \ln\,[(1 - J_2)\, Q_2 + J_2] \qquad \text{A4c(6)}$$

and

$$\frac{1}{Q_o - 1} = \frac{J_2}{Q_1 - 1} + \frac{1}{Q_2 - 1} \qquad \text{A4c(7)}$$

In addition for the stagewise case equations 4(15) and (17) become

$$\frac{(1 - J_1)\, Q_1 + J_1}{(Q_1 - 1)^2}\,\frac{\ln 1/J_1}{1 - J_1} = \frac{(1 - J_2)\, Q_2 + J_2}{J_2(Q_2 - 1)^2}\,\frac{\ln 1/J_2}{1 - J_2} \qquad \text{A4c(8)}$$

and

$$\frac{Q_1 - 1 - N_1\,((1 - J_1)\, Q_1 + J_1)}{(1 - J_1)\,(Q_1 - 1)^2}$$

$$= \frac{Q_2 - 1 - N_2((1 - J_2)\, Q_2 + J_2)/J_2}{(1 - J_2)\,(Q_2 - 1)^2} \qquad \text{A4c(9)}$$

For the differential case equations 4(16) and (18) become

$$\frac{(1 - J_1)\, Q_1 + J_1}{J_1(Q_1 - 1)^2} = \frac{(1 - J_2)\, Q_2 + J_2}{J_2(Q_2 - 1)^2} \qquad \text{A4c(10)}$$

and
$$\frac{Q_1 - 1 - N_1\,((1 - J_1)\,Q_1 + J_1)/J_1}{(1 - J_1)\,(Q_1 - 1)^2}$$

$$= \frac{Q_2 - 1 - N_2\,((1 - J_2)\,Q_2 + J_2)}{(1 - J_2)\,(Q_2 - 1)^2} \qquad \text{A4c(11)}$$

Inspection of equations A4c(5), (6), (7), (10) and (11) for the differential case shows that they are satisfied by $N_{L1}/J_1 = N_{G2}, = N_{TO}$, $J_1 = J_2$ and $Q_1 = Q_2$ as are the set of equations based on $Q_{3/4}$ in Appendix 4h. For the stagewise case the solution to equations A4c(5), (6), (7), (8) and (9) is similar to that of the set of equations based in $Q_{3/4}$ in Appendix 4h, and shows that for practical purposes $N_{S1} = N_{S2} = N_{SO}$. Graphs relating N_{SO} or J_2 to J_1J_2 and the overall separation factor $Q_{2/4}$ may be prepared in a similar way to Figs. 4.4 and 4.5 involving $Q_{3/4}$.

Note that when equilibrium is approached at the x inlet, $y_{out} - y(x_{in})$ is a minimum and so the overall separation factor $Q_{2/4}$ defined by equation 4c(4) is a maximum. Inversion of $Q_{2/4}$ gives

$$Q_{2/3} = \frac{y_{in} - y(x_{in})}{y_{in} - y(x_{out})} \qquad \text{A4c(12)}$$

so that when equilibrium is approached at the y inlet the overall separation factor $Q_{2/3}$ is also a maximum. The overall separation factor Q_o is thus always a maximum providing that J_1J_2 or its inverse $1/J_1J_2$ are made less than unity. Its behaviour in this respect is similar to that of the overall separation factor $Q_{3/4}$ and the inverse $Q_{4/3}$.

APPENDIX 4d DERIVATION OF $\left(\frac{\partial Q_o}{\partial N_1}\right)_{J_1}$ AND $\left(\frac{\partial Q_o}{\partial J_1}\right)_{N_1}$ IN FORWARD AND BACK EXTRACTION

We wish to obtain

$$\left(\frac{\partial Q_o}{\partial N_1}\right)_{J_1} \quad \text{and} \quad \left(\frac{\partial Q_o}{\partial J_1}\right)_{N_1}$$

at constant

$$J_1J_2 = K$$

and

$$N_1 + N_2 = P$$

Thus when J_1 is constant so is J_2 and

$$dN_2/dN_1 = -1$$

Similarly when N_1 is constant so is N_2 and

$$dJ_2/dJ_1 = -J_2/J_1$$

From equation 4(10) it follows that

$$Q_o = f(Q_1, Q_2, J_2)$$

so at constant J_1 (and hence J_2),

$$\left(\frac{\partial Q_o}{\partial N_1}\right)_{J_1} = \left(\frac{\partial Q_o}{\partial Q_1}\right)_{Q_2}\left(\frac{\partial Q_1}{\partial N_1}\right)_{J_1} + \left(\frac{\partial Q_o}{\partial Q_2}\right)_{Q_1}\left(\frac{\partial Q_2}{\partial N_1}\right)_{J_1} \quad \text{A4d(1)}$$

and at constant N_1,

$$\left(\frac{\partial Q_o}{\partial J_1}\right)_{N_1} = \left(\frac{\partial Q_o}{\partial Q_1}\right)_{Q_2,J_2}\left(\frac{\partial Q_1}{\partial J_1}\right)_{N_1} + \left(\frac{\partial Q_o}{\partial Q_2}\right)_{Q_1,J_1}\left(\frac{\partial Q_2}{\partial J_1}\right)_{N_1} + \left(\frac{\partial Q_o}{\partial J_2}\right)_{Q_1,Q_2}\frac{dJ_2}{dJ_1} \quad \text{A4d(2)}$$

Using equation 4(10)

$$\left(\frac{\partial Q_o}{\partial Q_1}\right)_{Q_2,J_2} = \frac{J_2 - J_1J_2}{1 - J_1J_2}\left(\frac{Q_o - 1}{Q_1 - 1}\right)^2 \quad \text{A4d(3)}$$

$$\left(\frac{\partial Q_o}{\partial Q_2}\right)_{Q_1,J_2} = \frac{1 - J_2}{1 - J_1J_2}\left(\frac{Q_o - 1}{Q_2 - 1}\right)^2 \quad \text{A4d(4)}$$

and

$$\left(\frac{\partial Q_o}{\partial J_2}\right)_{Q_1,Q_2} = \frac{(Q_o - 1)^2}{1 - J_1J_2}\left(\frac{1}{Q_2 - 1} - \frac{1}{Q_1 - 1}\right) \quad \text{A4d(5)}$$

Using equation 4(2)

$$\left(\frac{\partial Q_1}{\partial N_1}\right)_{J_1} = Q_1 \ln 1/J_1 \quad \text{A4d(6)}$$

for the stagewise case, and

$$\left(\frac{\partial Q_1}{\partial N_1}\right)_{J_1} = Q_1 (1 - J_1)/J_1 \quad \text{A4d(7)}$$

for the differential case.

Now

$$\left(\frac{\partial Q_2}{\partial N_1}\right)_{J_1} = \left(\frac{\partial Q_2}{\partial N_2}\right)_{J_2}\frac{dN_2}{dN_1} = -\left(\frac{\partial Q_2}{\partial N_2}\right)_{J_2} \quad \text{A4d(8)}$$

so using equation 4(6)

$$\left(\frac{\partial Q_2}{\partial N_1}\right)_{J_1} = -Q_2 \ln 1/J_2 \qquad \text{A4d(9)}$$

for the stagewise case, and

$$\left(\frac{\partial Q_2}{\partial N_1}\right)_{J_1} = -Q_2\,(1 - J_2) \qquad \text{A4d(10)}$$

for the differential case. Substituting equations A4d(3), (4), (6) and (9) into equation A4d(1) gives for the stagewise case

$$\left(\frac{\partial Q_o}{\partial N_1}\right)_{J_1} = \frac{(Q_o - 1)^2}{1 - J_1J_2}\left(\frac{Q_1J_2\,(1 - J_1)\ln 1/J_1}{(Q_1 - 1)^2} - \frac{Q_2\,(1 - J_2)\ln 1/J_2}{(Q_2 - 1)^2}\right) \qquad \text{A4d(11)}$$

and substituting equations A4d(3), (4), (7) and (10) into equation A4d(1) gives for the differential case

$$\left(\frac{\partial Q_o}{\partial N_1}\right)_{J_1} = \frac{(Q_o - 1)^2}{1 - J_1J_2}\left(Q_1\,\frac{(1 - J_1)^2}{(Q_1 - 1)^2}\,\frac{J_2}{J_1} - Q_2\,\frac{(1 - J_2)^2}{(Q_2 - 1)^2}\right) \qquad \text{A4d(12)}$$

Using equation 4(2)

$$\left(\frac{\partial Q_1}{\partial J_1}\right)_{N_1} = -\frac{N_1Q_1}{J_1} \qquad \text{A4d(13)}$$

for the stagewise case, and

$$\left(\frac{\partial Q_1}{\partial J_1}\right)_{N_1} = -\frac{N_1Q_1}{J_1^2} \qquad \text{A4d(14)}$$

for the differential case.

Now

$$\left(\frac{\partial Q_2}{\partial J_1}\right)_{N_1} = \left(\frac{\partial Q_2}{\partial J_2}\right)_{N_2}\frac{dJ_2}{dJ_1} = -\frac{J_2}{J_1}\left(\frac{\partial Q_2}{\partial J_2}\right) \qquad \text{A4d(15)}$$

so using equation 4(6)

$$\left(\frac{\partial Q_2}{\partial J_1}\right)_{N_1} = \frac{N_2Q_2}{J_1} \qquad \text{A4d(16)}$$

for the stagewise case, and

$$\left(\frac{\partial Q_2}{\partial J_1}\right)_{N_1} = \frac{J_2}{J_1}\,N_2Q_2 \qquad \text{A4d(17)}$$

for the differential case.

Substituting equations A4d(3), (4), (5), (13) and (16) into equation A4d(2) gives for the stagewise case,

$$\left(\frac{\partial Q_o}{\partial J_1}\right)_{N_1} = \frac{(Q_o - 1)^2}{1 - J_1 J_2}\frac{J_2}{J_1} \times$$

$$\times \left(\frac{Q_1 - 1 - N_1 Q_1 (1 - J_1)}{(Q_1 - 1)^2} - \frac{Q_2 - 1 - N_2 Q_2 (1 - J_2)/J_2}{(Q_2 - 1)^2}\right)$$

A4d(18)

Substituting equations A4d(3), (4), (5), (14) and (17) into equation A4d(2) gives for the differential case,

$$\left(\frac{\partial Q_o}{\partial J_1}\right)_{N_1} = \frac{(Q_o - 1)^2}{1 - J_1 J_2}\frac{J_2}{J_1} \times$$

$$\times \left(\frac{Q_1 - 1 - N_1 Q_1 (1 - J_1)/J_1}{(Q_1 - 1)^2} - \frac{Q_2 - 1 - N_2 Q_2 (1 - J_2)}{(Q_2 - 1)^2}\right)$$

A4d(19)

Setting $(\partial Q_o/\partial N_1)_{J_1}$ and $(\partial Q_o/\partial J_1)_{N_1}$ equal to zero for both the stagewise and differential cases gives equations 4(15), (16), (17) and (18). Alternatively the partial derivatives could be obtained using the technique of Lagrangian multipliers[44,45].

APPENDIX 4e DERIVATION OF $\left(\frac{\partial^2 Q_o}{\partial N_1^2}\right)_{J_1}$ AND $\frac{\partial^2 Q_o}{\partial N_1 \, \partial J_1}$ IN FORWARD AND BACK EXTRACTION

It follows from equations A4d(11) and (12) that

$$\left(\frac{\partial Q_o}{\partial N_1}\right)_{J_1} = f(J_1, J_2, Q_1, Q_2, Q_o) \qquad \text{A4e(0)}$$

so at constant J_1 (and hence J_2)

$$\left(\frac{\partial^2 Q_o}{\partial N_1^2}\right)_{J_1} = \frac{\partial f}{\partial Q_1}\left(\frac{\partial Q_1}{\partial N_1}\right)_{J_1} + \frac{\partial f}{\partial Q_2}\left(\frac{\partial Q_2}{\partial N_1}\right)_{J_1} + \frac{\partial f}{\partial Q_o}\left(\frac{\partial Q_o}{\partial N_1}\right)_{J_1} \qquad \text{A4e(1)}$$

and at constant N_1 (and hence N_2)

$$\frac{\partial^2 Q_o}{\partial N_1 \partial J_1} = \frac{\partial f}{\partial J_1} + \frac{\partial f}{\partial J_2}\frac{dJ_2}{dJ_1} + \frac{\partial f}{\partial Q_1}\left(\frac{\partial Q_1}{\partial J_1}\right)_{N_1} + \frac{\partial f}{\partial Q_2}\left(\frac{\partial Q_2}{\partial J_1}\right)_{N_1} + \frac{\partial f}{\partial Q_o}\left(\frac{\partial Q_o}{\partial J_1}\right)_{N_1} \qquad \text{A4e(2)}$$

For clarity the subscripts indicating the constant parameters have been omitted from these derivatives involving f; these follow directly from equation A 4e (0). When $(\partial Q_o/\partial N_1)_{J_1} = 0$ it follows that $\partial f/\partial Q_o = 0$, and so the last term in each equation disappears.

Using equation A4d(11) for the stagewise case

$$\frac{\partial f}{\partial Q_1} = -\left(\frac{Q_o - 1}{Q_1 - 1}\right)^2 \frac{J_2(1 - J_1)\ln 1/J_1}{1 - J_1J_2}\left(\frac{Q_1 + 1}{Q_1 - 1}\right) \qquad \text{A4e(3)}$$

$$\frac{\partial f}{\partial Q_2} = \left(\frac{Q_o - 1}{Q_2 - 1}\right)^2 \frac{(1 - J_2)\ln 1/J_2}{1 - J_1J_2}\left(\frac{Q_2 + 1}{Q_2 - 1}\right) \qquad \text{A4e(4)}$$

$$\frac{\partial f}{\partial J_1} = -\left(\frac{Q_o - 1}{Q_1 - 1}\right)\frac{J_2Q_1}{1 - J_1J_2}\left(\ln 1/J_1 + \frac{1 - J_1}{J_1}\right) \qquad \text{A4e(5)}$$

and

$$\frac{\partial f}{\partial J_2} = \left(\frac{Q_o - 1}{Q_2 - 1}\right)^2 \frac{Q_2}{1 - J_1J_2}\left(\ln 1/J_2 + \frac{1 - J_2}{J_2}\right) + \left(\frac{Q_o - 1}{Q_1 - 1}\right)^2 \frac{(1 - J_1)Q_1 \ln 1/J_1}{1 - J_1J_2} \qquad \text{A4e(6)}$$

Substituting equations A4e(3) and (4) together with equations A4d(6) and (9) into equation A4e(1) and using equation 4(15) leads to, for the stagewise case,

$$\left(\frac{\partial^2 Q_o}{\partial N_1^2}\right) = -\left(\frac{Q_o - 1}{Q_2 - 1}\right)^2 \frac{(1 - J_2)Q_2 \ln 1/J_2}{1 - J_1J_2} \times \left(\frac{Q_1 + 1}{Q_1 - 1}\ln 1/J_1 + \frac{Q_2 + 1}{Q_2 - 1}\ln 1/J_2\right) \qquad \text{A4e(7)}$$

It follows from equations 4(2) and 4 (6) that for positive N_1 or N_2 when J_1 or J_2 is less than one, Q_1 or Q_2 must be greater than one and vice versa. Thus $(\partial^2 Q_o/\partial N_1^2)_{J_1}$ is negative when J_1J_2 is less than one.

Substituting equations A 4e(3), (4), (5) and (6) together with equations A 4d(13) and (16) into equation A 4e(2) leads to, for the stagewise case,

$$\frac{\partial^2 Q_o}{\partial N_1 \partial J_1} = \left(\frac{Q_o - 1}{Q_1 - 1}\right)^2 \frac{J_2}{J_1} \frac{Q_1}{1 - J_1 J_2} \times$$

$$\times \left(N_1(1 - J_1) \ln 1/J_1 \frac{Q_1 + 1}{Q_1 - 1} - (1 - J_1) - \ln 1/J_1\right)$$

$$+ \left(\frac{Q_o - 1}{Q_2 - 1}\right)^2 \frac{J_2}{J_1} \frac{Q_2}{1 - J_1 J_2} \times$$

$$\times \left(N_2 \frac{(1 - J_2)}{J_2} \ln 1/J_2 \frac{Q_2 + 1}{Q_2 - 1} - \frac{(1 - J_2)}{J_2} - \ln 1/J_2\right) \quad \text{A 4e(8)}$$

Using equation A 4d(12) for the differential case

$$\frac{\partial f}{\partial Q_1} = -\left(\frac{Q_o - 1}{Q_1 - 1}\right)^2 \frac{(1 - J_1)^2}{1 - J_1 J_2} \frac{J_2}{J_1} \frac{Q_1 + 1}{Q_1 - 1} \quad \text{A 4e(9)}$$

$$\frac{\partial f}{\partial Q_2} = \left(\frac{Q_o - 1}{Q_2 - 1}\right)^2 \frac{(1 - J_2)^2}{1 - J_1 J_2} \frac{Q_2 + 1}{Q_2 - 1} \quad \text{A 4e(10)}$$

$$\frac{\partial f}{\partial J_1} = -\left(\frac{Q_o - 1}{Q_1 - 1}\right)^2 \frac{Q_2 J_2}{1 - J_1 J_2} \frac{1 - J_1^2}{J_1^2} \quad \text{A 4e(11)}$$

and

$$\frac{\partial f}{\partial J_2} = \frac{(Q_o - 1)^2}{1 - J_1 J_2} \left(\frac{Q_1}{(Q_1 - 1)^2} \frac{(1 - J_1)^2}{J_1} + \frac{2Q_2 (1 - J_2)}{(Q_2 - 1)^2}\right) \quad \text{A 4e(12)}$$

Substituting equations A 4e(9) and (10) together with equations A 4d(7) and (10) into equation A 4e(1) and using equation 4(16) leads to, for the differential case,

$$\left(\frac{\partial^2 Q_o}{\partial N_1^2}\right)_{J_1} = -\left(\frac{Q_o - 1}{Q_2 - 1}\right)^2 \frac{Q_2(1 - J_2)^2}{1 - J_1 J_2} \times$$

$$\times \left(\frac{Q_1 + 1}{Q_1 - 1} \frac{(1 - J_1)}{J_1} + \frac{Q_2 + 1}{Q_2 - 1} (1 - J_2)\right) \quad \text{A 4e(13)}$$

Substituting equations A4e(9), (10), (11) and (12) together with equations A4d(14) and (17) into equation A4e(2) leads to for the differential case,

$$\frac{\partial^2 Q_o}{\partial N_1 \partial J_1} = \left(\frac{Q_o - 1}{Q_1 - 1}\right)^2 \frac{(1 - J_1) Q_1}{1 - J_1 J_2} \frac{J_2}{J_1^2} \left(\frac{N_1}{J_1}(1 - J_1)\frac{Q_1 + 1}{Q_1 - 1} - 2\right)$$

$$+ \left(\frac{Q_o - 1}{Q_2 - 1}\right)^2 \frac{(1 - J_2) Q_2}{1 - J_1 J_2} \frac{J_2}{J_1} \left(N_2 (1 - J_2)\frac{Q_2 + 1}{Q_2 - 1} - 2\right)$$

A4e(14)

For the differential case, when $J_1 = J_2$; $N_1/J_1 = N_2$ and $Q_1 = Q_2$ equations A4e(13) and (14) become respectively,

$$\left(\frac{\partial^2 Q_o}{\partial N_1^2}\right)_{J_1} = -\left(\frac{Q_o - 1}{Q_1 - 1}\right)^2 \frac{Q_1 + 1}{Q_1 - 1} \frac{Q_1}{J_1} (1 - J_1)^2$$

(which is negative when Q_1 is greater than 1) and,

$$\frac{\partial^2 Q_o}{\partial N_1 \partial J_1} = \frac{Q_o - 1}{Q_1 - 1} \frac{Q_1}{J_1} \left(\frac{N_1}{J_1}(1 - J_1)\frac{Q_1 + 1}{Q_1 - 1} - 2\right)$$

APPENDIX 4f
DERIVATION OF $\left(\frac{\partial^2 Q_o}{\partial J_1^2}\right)_{N_1}$ AND $\frac{\partial^2 Q_o}{\partial J_1 \partial N_1}$ IN FORWARD AND BACK EXTRACTION

It follows from equations A4d(18) and (19) that

$$\left(\frac{\partial Q_o}{\partial J_1}\right)_{N_1} = f(J_1, J_2, N_1, N_2, Q_1, Q_2 \text{ and } Q_o) \qquad \text{A4f(0)}$$

so at constant N_1 (and hence N_2),

$$\left(\frac{\partial^2 Q_o}{\partial J_1^2}\right)_{N_1} = \frac{\partial f}{\partial J_1} + \frac{\partial f}{\partial J_2}\frac{dJ_2}{dJ_1} + \frac{\partial f}{\partial Q_1}\left(\frac{\partial Q_1}{\partial J_1}\right)_{N_1} + \frac{\partial f}{\partial Q_2}\left(\frac{\partial Q_2}{\partial J_1}\right)_{N_1}$$

A4f(1)

and at constant J_1 (and hence J_2),

$$\frac{\partial^2 Q_o}{\partial J_1 \partial N_1} = \frac{\partial f}{\partial N_1} + \frac{\partial f}{\partial N_2}\frac{dN_2}{dN_1} + \frac{\partial f}{\partial Q_1}\left(\frac{\partial Q_1}{\partial N_1}\right)_{J_1} + \frac{\partial f}{\partial Q_2}\left(\frac{\partial Q_2}{\partial N_1}\right)_{J_1}$$

A4f(2)

since $\dfrac{\partial f}{\partial Q_o} = 0$ when $\left(\dfrac{\partial Q_o}{\partial J_1}\right)_{N_1} = 0$. For clarity, in those derivatives involving f, the subscripts indicating the constant parameters have been omitted; these follow directly from equation A 4f(0).

Using equation A 4d(18) for the stagewise case,

$$\frac{\partial f}{\partial J_1} = \left(\frac{Q_o - 1}{Q_1 - 1}\right)^2 \frac{J_2}{J_1} \frac{N_1 Q_1}{1 - J_1 J_2} \qquad \text{A 4f(3)}$$

$$\frac{\partial f}{\partial J_2} = -\left(\frac{Q_o - 1}{Q_2 - 1}\right)^2 \frac{1}{J_1 J_2} \frac{N_2 Q_2}{1 - J_1 J_2} \qquad \text{A 4f(4)}$$

$$\frac{\partial f}{\partial Q_1} = -\left(\frac{Q_o - 1}{Q_1 - 1}\right)^2 \frac{J_2}{J_1 (1 - J_1 J_2)} \left(1 - N_1 (1 - J_1) \frac{Q_1 + 1}{Q_1 - 1}\right) \qquad \text{A 4f(5)}$$

$$\frac{\partial f}{\partial Q_2} = \left(\frac{Q_o - 1}{Q_2 - 1}\right)^2 \frac{J_2}{J_1 (1 - J_1 J_2)} \left(1 - N_2 \frac{(1 - J_2)}{J_2} \frac{Q_2 + 1}{Q_2 - 1}\right) \qquad \text{A 4f(6)}$$

$$\frac{\partial f}{\partial N_1} = \left(\frac{Q_o - 1}{Q_1 - 1}\right)^2 \frac{J_2}{J_1} \frac{(1 - J_1) Q_1}{1 - J_1 J_2} \qquad \text{A 4f(7)}$$

$$\frac{\partial f}{\partial N_2} = \left(\frac{Q_o - 1}{Q_2 - 1}\right)^2 \frac{1}{J_1} \frac{(1 - J_2) Q_2}{1 - J_1 J_2} \qquad \text{A 4f(8)}$$

Substituting equations A 4f(3), (4) and (5) and (6) together with equations A 4d(13) and (16) into equation A 4f(1), leads to, for the stagewise case,

$$\left(\frac{\partial^2 Q_o}{\partial J_1^2}\right)_{N_1} = \left(\frac{Q_o - 1}{Q_1 - 1}\right)^2 \frac{J_2}{J_1^2} \frac{N_1 Q_1}{1 - J_1 J_2} \times$$
$$\times \left(1 + J_1 - N_1 (1 - J_1) \frac{Q_1 + 1}{Q_1 - 1}\right)$$
$$+ \left(\frac{Q_o - 1}{Q_2 - 1}\right)^2 \frac{1}{J_1^2} \frac{N_2 Q_2}{1 - J_1 J_2} \times$$
$$\times \left(1 + J_2 - N_2 (1 - J_2) \frac{Q_2 + 1}{Q_2 - 1}\right) \qquad \text{A 4f(9)}$$

This is negative when J_1J_2 is less than unity and the last term in each set of *large* brackets is greater than the first or vice versa. Consideration of equations 4(2) and 4(6) shows the last terms are always positive.

Substituting equations A4f(5), (6), (7) and (8) together with equations A4d(6) and (9) into equation A4f(2) and collecting terms leads to, for the stagewise case,

$$\frac{\partial^2 Q_o}{\partial J_1\,\partial N_1} = \left(\frac{Q_o - 1}{Q_1 - 1}\right)^2 \frac{J_2}{J_1} \frac{Q_1}{1 - J_1J_2} \times$$
$$\times \left(N_1(1 - J_1)\frac{Q_1 + 1}{Q_1 - 1}\ln 1/J_1 - \ln 1/J_1 - (1 - J_1)\right)$$
$$+ \left(\frac{Q_o - 1}{Q_2 - 1}\right)^2 \frac{J_2}{J_1} \frac{Q_2}{1 - J_1J_2} \times$$
$$\times \left(N_2 \frac{(1 - J_2)}{J_2} \frac{Q_2 + 1}{Q_2 - 1}\ln 1/J_2 - \ln 1/J_2 - \frac{(1 - J_2)}{J_2}\right) \qquad \text{A4f(10)}$$

which is identical with $\dfrac{\partial^2 Q_o}{\partial N_1\,\partial J_1}$ given by equation A4c(8).

For the differential case using equation A4d(19)

$$\frac{\partial f}{\partial J_1} = \left(\frac{Q_o - 1}{Q_1 - 1}\right)^2 \frac{J_2}{J_1^3} \frac{N_1Q_1}{1 - J_1J_2} \qquad \text{A4f(11)}$$

$$\frac{\partial f}{\partial J_2} = -\left(\frac{Q_o - 1}{Q_2 - 1}\right)^2 \frac{J_2}{J_1} \frac{N_2Q_2}{1 - J_1J_2} \qquad \text{A4f(12)}$$

$$\frac{\partial f}{\partial Q_1} = -\left(\frac{Q_o - 1}{Q_1 - 1}\right)^2 \frac{J_2}{J_1} \frac{1}{1 - J_1J_2}\left(1 - \frac{N_1}{J_1}(1 - J_1)\frac{Q_1 + 1}{Q_1 - 1}\right) \qquad \text{A4f(13)}$$

$$\frac{\partial f}{\partial Q_2} = \left(\frac{Q_o - 1}{Q_2 - 1}\right)^2 \frac{J_2}{J_1} \frac{1}{1 - J_1J_2}\left(1 - N_2(1 - J_2)\frac{Q_2 + 1}{Q_2 - 1}\right) \qquad \text{A4f(14)}$$

$$\frac{\partial f}{\partial N_1} = -\left(\frac{Q_o - 1}{Q_1 - 1}\right)^2 \frac{J_2}{J_1^2} \frac{(1 - J_1)\,Q_1}{1 - J_1J_2} \qquad \text{A4f(15)}$$

and

$$\frac{\partial f}{\partial N_2} = \left(\frac{Q_o - 1}{Q_2 - 1}\right)^2 \frac{J_2}{J_1} \frac{(1 - J_2) Q_2}{1 - J_1 J_2} \qquad \text{A4f(16)}$$

Substituting equations A4f(11), (12), (13) and (14) together with equations A4d(14) and (17) into equation A4f(1) and collecting terms yields, for the differential case,

$$\left(\frac{\partial^2 Q_o}{\partial J_1^2}\right)_{N_1} = \left(\frac{Q_o - 1}{Q_1 - 1}\right)^2 \frac{J_2}{J_1^3} \frac{N_1 Q_1}{1 - J_1 J_2} \left(2 - N_1 \frac{(1 - J_1)}{J_1} \frac{Q_1 + 1}{Q_1 - 1}\right)$$
$$+ \left(\frac{Q_o - 1}{Q_2 - 1}\right)^2 \frac{J_2^2}{J_1^2} \frac{N_2 Q_2}{1 - J_1 J_2} \left(2 - N_2 (1 - J_2) \frac{Q_2 + 1}{Q_2 - 1}\right) \qquad \text{A4f(17)}$$

Substituting equations A4f(13), (14), (15) and (16) together with equations A4d(7) and (10) into equation A4f(2) and collecting terms, yields, for the differential case,

$$\frac{\partial^2 Q_o}{\partial J_1 \partial N_1} = \left(\frac{Q_o - 1}{Q_1 - 1}\right)^2 \frac{J_2}{J_1^2} \frac{(1 - J_1) Q_1}{1 - J_1 J_2} \left(\frac{N_1}{J_1} (1 - J_1) \frac{Q_1 + 1}{Q_1 - 1} - 2\right)$$
$$+ \left(\frac{Q_o - 1}{Q_2 - 1}\right)^2 \frac{J_2}{J_1} \frac{(1 - J_2) Q_2}{1 - J_1 J_2} \left(N_2 (1 - J_2) \frac{Q_2 + 1}{Q_2 - 1} - 2\right). \qquad \text{A4f(18)}$$

which is identical with $\partial^2 Q_o/(\partial N_1 \partial J_1)$ given by equation A4e(14).

When $J_1 = J_2$, $N_1/J_1 = N_2$ and $Q_1 = Q_2$ equations A4f(17) and (18) for the differential case become,

$$\left(\frac{\partial^2 Q_o}{\partial J_1^2}\right)_{N_1} = \left(\frac{Q_o - 1}{Q_1 - 1}\right)^2 \frac{1}{J_1^2} \frac{N_1 Q_1}{1 - J_1} \left(2 - N_1 \frac{(1 - J_1)}{J_1} \frac{Q_1 + 1}{Q_1 - 1}\right)$$

and

$$\frac{\partial^2 Q_o}{\partial J_1 \partial N_1} = \left(\frac{Q_o - 1}{Q_1 - 1}\right)^2 \frac{Q_1}{J_1} \left(N_1 \frac{(1 - J_1)}{J_1} \frac{Q_1 + 1}{Q_1 - 1} - 2\right)$$

The former equation shows that $(\partial^2 Q_o / \partial J_1^2)_{N_1}$ is negative when $N_1 \dfrac{(1 - J_1)}{J_1} \dfrac{Q_1 + 1}{Q_1 - 1}$ is greater than 2 and J_1 is less than 1.

APPENDIX 4g
DERIVATION OF $\left(\frac{\partial P}{\partial N_1}\right)_{J_1}$ AND $\left(\frac{\partial P}{\partial J_1}\right)_{N_1}$ IN FORWARD AND BACK EXTRACTION

We wish to obtain

$$\left(\frac{\partial P}{\partial N_1}\right)_{J_1} \quad \text{and} \quad \left(\frac{\partial P}{\partial J_1}\right)_{N_1}$$

at constant $J_1 J_2$ and Q_o. Thus when J_1 is constant, so is J_2, and when N_1 is constant

$$\frac{dJ_2}{dJ_1} = -\frac{J_2}{J_1}$$

Equation 4(13) is

$$P = N_1 + N_2$$

so it follows that

$$\left(\frac{\partial P}{\partial N_1}\right)_{J_1} = 1 + \left(\frac{\partial N_2}{\partial N_1}\right)_{J_1}$$

$$= 1 + \left(\frac{\partial N_2}{\partial Q_2}\right)_{J_2}\left(\frac{\partial Q_2}{\partial Q_1}\right)_{J_2}\left(\frac{\partial Q_1}{\partial N_1}\right)_{J_1} \qquad \text{A4g(1)}$$

and

$$\left(\frac{\partial P}{\partial J_1}\right)_{N_1} = \left(\frac{\partial N_2}{\partial J_1}\right)_{N_1} = \frac{\partial N_2}{\partial J_2}\left(\frac{\partial J_2}{\partial J_1}\right)_{N_1} + \frac{\partial N_2}{\partial Q_2}\left(\frac{\partial Q_2}{\partial J_1}\right)_{N_1}$$

$$= -\frac{J_2}{J_1}\frac{\partial N_2}{\partial J_2} + \frac{\partial N_2}{\partial Q_2}\left(\frac{\partial Q_2}{\partial Q_1}\left(\frac{\partial Q_1}{\partial J_1}\right)_{N_1} + \frac{\partial Q_2}{\partial J_2}\left(\frac{\partial J_2}{\partial J_1}\right)_{N_1}\right) \qquad \text{A4g(2)}$$

for N_2 is a function of J_2 and Q_2 by equation 4(6) and Q_2 is a function of Q_1 and J_2 by equation 4(10). From the latter equation 4(10) at constant $J_1 J_2$ and Q_o it follows that

$$\frac{\partial Q_2}{\partial Q_1} = -J_2\frac{1-J_1}{1-J_2}\left(\frac{Q_2-1}{Q_1-1}\right)^2 \qquad \text{A4g(3)}$$

and

$$\frac{\partial Q_2}{\partial J_2} = -\frac{(Q_2-1)^2}{1-J_2}\left(\frac{1}{Q_1-1} - \frac{1}{Q_2-1}\right) \qquad \text{A4g(4)}$$

Using the former equation 4(6)

$$\frac{\partial N_2}{\partial J_2} = \frac{N_2}{J_2 \ln 1/J_2} \qquad \text{A4g(5)}$$

for the stagewise case, and

$$\frac{\partial N_2}{\partial J_2} = \frac{N_2}{1 - J_2} \qquad \text{A4g(6)}$$

for the differential case.

Substituting equation A4g(3) together with equations A4d(6), (8) and (9) into equation A4g(1) gives for the stagewise case,

$$\left(\frac{\partial P}{\partial N_1}\right)_{J_1} = 1 - \frac{Q_1}{Q_2} J_2 \frac{1 - J_1}{1 - J_2} \frac{\ln 1/J_1}{\ln 1/J_2} \left(\frac{Q_2 - 1}{Q_1 - 1}\right)^2 \qquad \text{A4g(7)}$$

and substituting A4g(3) together with equations A4d(7), (8) and (10) into equation A4g(1) gives for the differential case,

$$\left(\frac{\partial P}{\partial N_1}\right)_{J_1} = 1 - \frac{J_2}{J_1} \left(\frac{1 - J_1}{1 - J_2}\right)^2 \frac{Q_1}{Q_2} \left(\frac{Q_2 - 1}{Q_1 - 1}\right)^2 \qquad \text{A4g(8)}$$

Substituting equations A4g(3), (4) and (5) together with equations A4d(8), (9) and (13) into equation A4g(2) yields for the stagewise case,

$$\left(\frac{\partial P}{\partial J_1}\right)_{N_1} = \frac{J_2}{J_1} \frac{(Q_2 - 1)^2}{1 - J_2} \frac{1}{Q_2 \ln 1/J_2} \times$$

$$\times \left(\frac{(Q_2 - 1) - N_2 Q_2 (1 - J_2)/J_2}{(Q_2 - 1)^2} - \frac{(Q_1 - 1) - N_1 Q_1 (1 - J_1)}{(Q_1 - 1)^2}\right) \qquad \text{A4g(9)}$$

Substituting equations A4g(3), (4) and (6) together with equations A4d(8), (10) and (14) into equation A4g(2) yields for the differential case,

$$\left(\frac{\partial P}{\partial J_1}\right)_{N_1} = \frac{J_2}{J_1} \frac{(Q_2 - 1)^2}{Q_2 (1 - J_2)^2} \times$$

$$\times \left(\frac{(Q_2 - 1) - N_2 Q_2 (1 - J_2)}{(Q_2 - 1)^2} - \frac{(Q_1 - 1) - N_1 Q_1 (1 - J_1)/J_1}{(Q_1 - 1)^2}\right) \qquad \text{A4g(10)}$$

When

$$\left(\frac{\partial P}{\partial N_1}\right)_{J_1} = \left(\frac{\partial P}{\partial J_1}\right)_{N_1} = 0$$

equations A4g(7) and (9) for the stagewise case become equations 4(15) and 4(17) whilst equations A4g(8) and (10) for the differential case become equations 4(16) and 4(18).

The second-order differentials may be obtained in a similar manner to that described in Appendices A4e and A4f. As P is a minimum under optimum conditions it follows that

$$\left(\frac{\partial^2 P}{\partial N_1^2}\right)_{J_1} \quad \text{and} \quad \left(\frac{\partial^2 P}{\partial J_1^2}\right)_{N_1}$$

must both be positive and their product greater than

$$\left(\frac{\partial^2 P}{\partial N_1 \partial J_1}\right)^2.$$

APPENDIX 4h
SOLUTION OF THE EQUATIONS GOVERNING THE OPTIMISATION OF FORWARD AND BACK EXTRACTION IN A STAGEWISE CONTACTOR

The optimisation of forward and back extraction in a stagewise contactor is governed by equations 4(2), (6), (10), (12), (13), (15) and (17) which are listed respectively below,

$$N_{S1} \ln 1/J_1 = \ln Q_1 \qquad \text{A4h(1)}$$

$$N_{S2} \ln 1/J_2 = \ln Q_2 \qquad \text{A4h(2)}$$

$$\frac{1 - J_1 J_2}{Q_O - 1} = \frac{1 - J_2}{Q_2 - 1} + \frac{J_2 - J_1 J_2}{Q_1 - 1} \qquad \text{A4h(3)}$$

$$J_1 J_2 = K \qquad \text{A4h(4)}$$

$$N_1 + N_2 = P \qquad \text{A4h(5)}$$

$$\frac{(1 - J_1) Q_1}{(Q_1 - 1)^2} \ln 1/J_1 = \frac{(1 - J_2) Q_2}{J_2 (Q_2 - 1)^2} \ln 1/J_2 \qquad \text{A4h(6)}$$

and

$$\frac{Q_1 - 1 - N_1 (1 - J_1) Q_1}{(Q_1 - 1)^2} = \frac{Q_2 - 1 - N_2 (1 - J_2) Q_2/J_2}{(Q_2 - 1)^2}$$

A4h(7)

There are thus seven equations and nine variables J_1, J_2, Q_1, Q_2, N_1, N_2, K, P and Q_O. The problem is to find the ratio of the numbers of stages in contactors 1 and 2, N_1/N_2, and the solvent recirculation rate S (i.e. J_1 or J_2) such that the overall separation Q_o is a maximum, knowing the overall operating conditions L, G, m_1, m_2 (i.e. K) and the total number of stages, P. However, mathematically speaking it is sufficient to fix any two variables in order to obtain the other seven. The number of ways of choosing two from nine is

$$^2C_9 = \frac{9!}{2!\,7!} = 36$$

so there are this many alternative methods of solution. The one chosen will depend on its simplicity and the form of graphical presentation. Fortunately it transpires that the solution requires that for most practical purposes N_1 must equal N_2. Figure 4.3 representing the solution for the differential case thus suggests an analogous stagewise plot of log Q_O versus log K for different values of $N_1 = N_2 = N_{SO}$, say. Alternatively the differential solution equation 4(19) also suggests a stagewise plot of N_{SO} versus log Q_O for different values of K. Solution of the governing equations is simpler and physically more meaningful if K is kept constant rather than N_1 or N_2. Furthermore, it is convenient to also hold J_1 constant for this immediately fixes J_2. The mathematical problem is thus to obtain N_1, N_2, Q_1, Q_2, P, Q_O and J_2 knowing K and J_1.

Suppose we guess Q_1 and Q_2, then N_1 and N_2 will follow from equations A4h(1) and (2); Q_O and P also follow from equations A4h(3) and (5). The values of Q_1 and Q_2 may thus be refined using equations A4h(6) and (7). Let us define

$$A_1 = \frac{(1 - J_1)\,Q_1}{(Q_1 - 1)^2} \ln 1/J_1 \qquad \text{A4h(8)}$$

$$A_2 = \frac{(1 - J_2)\,Q_2}{J_2\,(Q_2 - 1)^2} \ln 1/J_2 \qquad \text{A4h(9)}$$

and

$$f = A_1 - A_2 \qquad \text{A4h(10)}$$

also

$$B_1 = \frac{Q_1 - 1 - N_1(1 - J_1)Q_1}{(Q_1 - 1)^2} \qquad \text{A4h(11)}$$

$$B_2 = \frac{Q_2 - 1 - N_2(1 - J_2)Q_2/J_2}{(Q_2 - 1)^2} \qquad \text{A4h(12)}$$

and

$$g = B_1 - B_2 \qquad \text{A4h(13)}$$

Dividing equations A4h(6) and (7) gives

$$\frac{N_1Q_1(1 - J_1) + 1 - Q_1}{(1 - J_1)Q_1 \ln 1/J_1} = \frac{N_2Q_2(1 - J_2) + (1 - Q_2)J_2}{(1 - J_2)Q_2 \ln 1/J_2} \qquad \text{A4h(14)}$$

so it is convenient also to define

$$C_1 = \frac{N_1}{\ln 1/J_1} + \frac{1/Q_1 - 1}{(1 - J_1)\ln 1/J_1} \qquad \text{A4h(15)}$$

$$C_2 = \frac{N_2}{\ln 1/J_2} + \frac{(1/Q_2 - 1)J_2}{(1 - J_2)\ln 1/J_2} \qquad \text{A4h(16)}$$

and

$$h = C_1 - C_2 \qquad \text{A4h(17)}$$

Then using the Newton–Raphson procedure[47] better estimates of Q_1 and Q_2 (denoted Q_1' and Q_2') are

$$Q_1' = Q_1 + eQ_1 \qquad \text{A4h(18)}$$

and

$$Q_2' = Q_2 + eQ_2 \qquad \text{A4h(19)}$$

where eQ_1 and eQ_2 are estimates of the errors in the values of Q_1 and Q_2 given by

$$eQ_1\left(\frac{\partial f}{\partial Q_1}\right)_{Q_2} + eQ_2\left(\frac{\partial f}{\partial Q_2}\right)_{Q_1} = -f \qquad \text{A4h(20)}$$

and

$$eQ_1\left(\frac{\partial h}{\partial Q_1}\right)_{Q_2} + eQ_2\left(\frac{\partial h}{\partial Q_2}\right)_{Q_1} = -h \qquad \text{A4h(21)}$$

where the values of f and h and their derivatives are at constant J_1 and J_2 and are evaluated at Q_1 and Q_2. (These equations employ the function h rather than g as the derivatives of h are simpler than those of g.)

Solving the simultaneous equations yields

$$eQ_1 = \left(-f\frac{\partial h}{\partial Q_2} + h\frac{\partial f}{\partial Q_2}\right)\Big/\left(\frac{\partial f}{\partial Q_1}\frac{\partial h}{\partial Q_2} - \frac{\partial f}{\partial Q_2}\frac{\partial h}{\partial Q_1}\right) \quad \text{A 4h(22)}$$

$$eQ_2 = \left(-h\frac{\partial f}{\partial Q_1} + f\frac{\partial h}{\partial Q_1}\right)\Big/\left(\frac{\partial f}{\partial Q_1}\frac{\partial h}{\partial Q_2} - \frac{\partial f}{\partial Q_2}\frac{\partial h}{\partial Q_1}\right) \quad \text{A 4h(23)}$$

which enable the second estimates Q_1' and Q_2' to be obtained. These may be further refined in similar fashion and the iteration repeated until two successive estimates of Q_1 or Q_2 agree to within some specified value, say 0·1%. To avoid over-correction it is advantageous to rewrite equations 4h(18) and (19)

$$Q_1' = Q_1 + eQ_1/S \quad \text{A 4h(24)}$$

$$Q_2' = Q_2 + eQ_2/S \quad \text{A 4h(25)}$$

where S is a number greater than one. By varying the step length S it is possible to obtain acceptable values of Q_1 and Q_2 with the minimum number of iterations. The iterative method is sensitive to the initial guesses of Q_1 and Q_2 but careful choice of S overcomes this sensitivity to some extent.

Differentiating equation 4h(10) with respect to Q_1 and Q_2 at constant J_1 and J_2 using equations A 4h(8) and (9) gives

$$\left(\frac{\partial f}{\partial Q_1}\right)_{Q_2} = -\frac{Q_1 + 1}{(Q_1 - 1)^3}(1 - J_1)\ln 1/J_1 \quad \text{A 4h(26)}$$

and

$$\left(\frac{\partial f}{\partial Q_2}\right)_{Q_1} = \frac{Q_2 + 1}{(Q_2 - 1)^3}\frac{1 - J_2}{J_2}\ln 1/J_2 \quad \text{A 4h(27)}$$

Differentiating equation 4h(17) with respect to Q_1 and Q_2 at constant J_1 and J_2 using equations A 4h(15) and (16) gives

$$\left(\frac{\partial h}{\partial Q_1}\right)_{Q_2} = \frac{1}{\ln 1/J_1}\left(\frac{\partial N_1}{\partial Q_1}\right)_{J_1} - \frac{1/Q_1}{(1 - J_1)\ln 1/J_1} \quad \text{A 4h(28)}$$

and

$$\left(\frac{\partial h}{\partial Q_2}\right)_{Q_1} = -\frac{1}{\ln 1/J_2}\left(\frac{\partial N_2}{\partial Q_2}\right)_{J_2} + \frac{J_2/Q_2}{(1 - J_2)\ln 1/J_2} \quad \text{A 4h(29)}$$

It follows from equations A 4h(1) and (2) that

$$\left(\frac{\partial N_1}{\partial Q_1}\right)_{J_1} = \frac{1/Q_1}{\ln 1/J_1} \qquad \text{A 4h(30)}$$

and

$$\left(\frac{\partial N_2}{\partial Q_2}\right)_{J_2} = \frac{1/Q_2}{\ln 1/J_2} \qquad \text{A 4h(31)}$$

so equations A 4h(28) and (29) become

$$\frac{\partial h}{\partial Q_1} = \frac{1/Q_1}{\ln 1/J_1}\left(\frac{1}{\ln 1/J_1} - \frac{1/Q_1}{1 - J_1}\right) \qquad \text{A 4h(32)}$$

and

$$\frac{\partial h}{\partial Q_2} = -\frac{1/Q_2}{\ln 1/J_2}\left(\frac{1}{\ln 1/J_2} - \frac{J_2/Q_2}{1 - J_2}\right) \qquad \text{A 4h(33)}$$

The derivatives given by equations A 4h(26), (27), (32) and (33) may be substituted into equations A 4h(22) and (23) so that eQ_1 and eQ_2 may be evaluated.

The technique described above was used to solve equations A 4h(1) to (7) on a KDF 9 digital computer using Atlas autocode. A typical program Q16/15, with part of the output, is given in Appendix 4m, to illustrate the method, and graphical procedures are included. The program is intended to be self-explanatory rather than efficient. It includes the variables N_R, Q_R, and β which relate to cross-current extraction with recirculation discussed in Chapter 5. Results have also been obtained using a modification of the more efficient routine *Multifit*[(49)].

The results show that N_{S1} must almost exactly equal N_{S2} to get the maximum separation Q_O from a given $N_{S1} + N_{S2}$, but that the optimum recirculation rate S is not given when J_1 and J_2 are equal. N_{S1} and N_{S2} appear to diverge slightly (by about 1 %) when they are small (less than 2) and J_1J_2 is small (less than 0·1). Figures 4.4 and 4.5 are plots of $N_{S1} = N_{S2} = N_{SO}$ and J_2 versus Q_O for different values of J_1J_2.

It is also possible to guess N_1 and N_2 rather than Q_1 and Q_2 and to refine the initial values by a similar method to that described above. However, this alternative is not so stable for $Q_1 = (1/J_1)^{N1}$ and $Q_2 = (1/J_2)^{N2}$ and so exponential overflows may occur during the iteration when N_1 and N_2 are estimated. When Q_1 and Q_2 are estimated, fluctuations in their values during iteration do not cause instability because of the logarithmic dependence of N_1 and N_2 on Q_1 and Q_2 indicated by equations A 4h(1) and (2).

Another way of obtaining the values of Q_1 and Q_2 which satisfy equations A 4h(1) to (7) when J_1 and J_2 are fixed is to carry out a search over a wide range of values of Q_1, say. For each value of Q_1 the value of Q_2 may be obtained from equations A 4h(6) and hence N_1 and N_2 from equations A 4h(1) and (2). However, only those values of J_1, J_2, Q_1, Q_2, N_1 and N_2 which satisfy equation A 4h(7) are solutions to the general problem of optimisation of forward and back extraction in a stagewise contactor.

A better method is to eliminate Q_2 between equations A 4h(3) and (6) and obtain an equation relating Q_1 to J_1, J_2 and Q_O. It is thus possible to fix $J_1 J_2$ and Q_O and to carry out a search over a wide range of values of J_1 say. This satisfies the physical problem of finding the ratio of N_1/N_2 and the solvent recirculation rate S (i.e. J_1 or J_2) such that the total number of stages P is a minimum, for a given overall separation Q_O knowing the overall operating conditions L, G, m_1 and m_2 (i.e. $J_1 J_2$). Appendix 4g shows that this leads to a set of equations identical with equations A 4h(1) to (7).

Eliminating Q_2 between equations A 4h(3) and (6) gives the quadratic in $(Q_1 - 1)$

$$\left(\frac{Q_1 - 1}{Q_O - 1}\right)^2 \frac{1 - J_1 J_2}{1 - J_1} \frac{R_2}{R_1} \left(1 + \frac{(1 - J_2) Q_O}{(1 - J_1) J_2}\right)$$
$$- \left(\frac{Q_1 - 1}{Q_O - 1}\right)\left(\frac{R_2}{R_1} \frac{1 - J_2}{1 - J_1} Q_O + J_2 + \frac{1 - J_1 J_2}{1 - J_1} + Q_O - 1\right)$$
$$+ J_2 \frac{R_2}{R_1} - 1 = 0 \qquad \text{A 4h(34)}$$

where $R_1 = \dfrac{\ln 1/J_1}{1 - J_1}$ and $R_2 = \dfrac{\ln 1/J_2}{1 - J_2}$

Thus if $J_1 J_2$ and Q_O are fixed and a search is carried out over a wide range of values of J_1, it is possible to obtain two values of Q_1, for each value of J_1, from this equation. The corresponding values of Q_2

follow from equations A4h(3) or (6) and of N_1 and N_2 from equations A4h(1) and (2). However, as before only those values of J_1, J_2, Q_1, Q_2, N_1 and N_2 which satisfy equation A4h(7) are solutions to the general problem of optimisation of forward and back extraction in a stagewise contactor.

The values of B_1 and B_2 may be obtained from equations A4h(11) and (12) and hence the value of g from equation A4h(13). When g changes sign between successive values of J_1 these must straddle the exact value of J_1. It is thus possible to print out the values of J_1 and the other variables when g is close to zero. The method of solution is similar to the technique of using contour charts.[26]

A typical program 15/9 written in Atlas autocode for a KDF9 digital computer is given in Appendix 4n to illustrate the method; part of the output is also shown. Although more productive, the search technique is not as rigorous as the iterative method described above and results must be selected from the output with care. The program is intended to be self-explanatory rather than efficient. It includes the variables N_R, Q_R and β which relate to cross-current extraction with recirculation discussed in Chapter 5.

APPENDIX 4i
LIMITING CASES OF THE EQUATIONS GOVERNING FORWARD AND BACK EXTRACTION IN A STAGEWISE CONTACTOR

$J_1 J_2 = 1$

Writing $J_2 = 1/J_2 = 1/J_1$ shows that the governing equations 4(2), (6), (10), (15) and (17) are only satisfied by $J_1 = J_2 = 1$, $N_{S1} = N_{S2} = N_{SO}$ and $Q_1 = Q_2 = Q_O = 1$. However, if equation 4(10) is re-expressed in terms of $Q_{2/4}$ (defined by equation 4(29)) as in Appendix 4b, substituting $J_1 = J_2 = 1$ yields

$$N_{SO} = 2\,(Q_{2/4} - 1)$$

which is an exact result analogous to that for the differential case given by equation 4(30).

$J_1 = J_2 = J_1J_2$

Putting $J_1 = J_2$ in equations 4(2) and (6) shows that $Q_1 = Q_2$ when $N_{S1} = N_{S2} = N_{SO}$. It thus follows from equation 4(10) that $Q_1 = Q_2 = Q_O$ and so

$$N_{SO} \ln 1/\sqrt{(J_1J_2)} = \ln Q_O$$

which is identical with equation 4(31). This is the approximate result which applies to most of the practical range except when N_{SO} is small. The relationship becomes more exact as J_1J_2 approaches unity because it is only exactly true that $J_1 = J_2$ when $J_1J_2 = 1$, as indicated above.

Infinite recirculation $J_1 \to \infty, J_2 \to 0$

When the flowrate S of the intermediate solvent is infinite $J_2 = m_2G/S$ is zero and $J_1 = m_1S/L$ approaches infinity. However, the overall operating conditions $J_1J_2 = m_1m_2G/L$ are still finite, and if N_1 and N_2 are finite $Q_1 = (1/J_1)^{N_1}$ is zero and $Q_2 = (1/J_2)^{N_2}$ approaches infinity. Substitution in equation 4(10) shows that the overall separation $Q_O = 1/J_1J_2$, which is reflected in Fig. 4.5.

Zero recirculation $J_1 \to 0, J_2 \to \infty$

When the flowrate S of the intermediate solvent is zero, $J_2 = m_2G/S$ approaches infinity and $J_1 = m_1S/L$ is zero, the overall operating conditions J_1J_2 being finite. If N_1 and N_2 are finite Q_2 is zero and Q_1 approaches infinity. Equation 4(10) shows that $Q_O = 1$ and so there is no overall separation, as must be expected for the intermediate solvent is transferring no solute.

$J_1 \to 1, J_2 \to J_1J_2$

The results of the computer program show that as N_1 and N_2 become smaller so J_1 approaches 1 and J_2 approaches J_1J_2. Writing $J_1 = 1 + \delta$ and $Q_1 = (1 - \delta)^{N_1}$, expanding and letting δ approach zero, shows that A_1 and B_1 (defined in Appendix 4h) reduce to $1/N_1^2$ and $(1 - N_1)/2N_1$ respectively, so equations 4(15) and (17) may be written

$$N_1^2 = \frac{J_2}{(1 - J_2) \ln 1/J_2}\left(Q_2 - 2 + \frac{1}{Q_2}\right)$$

and

$$\frac{N_1(1 - N_1)}{2} = \frac{J_2}{(1 - J_2) \ln 1/J_2}\left(1 - \frac{1}{Q_2} - N_2\frac{(1 - J_2)}{J_2}\right)$$

In addition equation 4(10) becomes

$$\frac{1 - J_1J_2}{Q_O - 1} = \frac{1 - J_1J_2}{Q_1 - 1} + \frac{J_1J_2}{N_1}$$

and these equations are satisfied by $Q_1 = Q_O = 1$ and $N_1 = N_2 = 0$ which indicates that the lines of constant J_1J_2 pass through the origin of Fig. 4.4.

$J_1 \to J_1J_2$, $J_2 \to 1$

The results of the computer program show that as N_1 and N_2 become larger so J_1 approaches J_1J_2 and J_2 approaches unity. Writing $J_2 = 1 + \delta$ in equation 4(15) shows that Q_1 approaches infinity and using this fact in equation 4(17) shows that Q_2 must also approach infinity. It follows from equation 4(10) that Q_O tends to infinity which is indicated by Fig. 4.4 when N_1 and N_2 tend to infinity.

APPENDIX 4j THE OPTIMUM OPERATING CONDITIONS IN AN EXISTING DIFFERENTIAL FORWARD AND BACK EXTRACTOR

The optimum operation of an existing differential forward and back extractor is governed by equations 4(45) and (46)

$$\left(\frac{J_1J_2}{N_{L1}} + \frac{1}{N_{G2}}\right) \ln Q_O = 1 - J_1J_2$$

and

$$\left(1 - \frac{\ln Q_O}{N_{L1}/J_1}\right)\left(1 - \frac{\ln Q_O}{N_{G2}}\right) = J_1J_2$$

Figures A4j.1 to 10 give the variation of N_{L1} and N_{L1}/J_1 with N_{G2} at different values of Q_O for $J_1J_2 = 0{\cdot}1$, 0·3, 0·5, 0·7 and 0·9. Knowing the numbers of transfer units N_{L1} and N_{G2} in contactors 1 and 2 and the overall operating conditions J_1J_2 ($= m_1m_2G/L$) enables the maximum overall separation Q_O to be obtained from the odd-numbered figures. The value of N_{L1}/J_1 may then be obtained from the corresponding even-numbered figures which gives J_1 ($= m_1S/L$) and hence the optimum value of the recirculation rate for the intermediate solvent S.

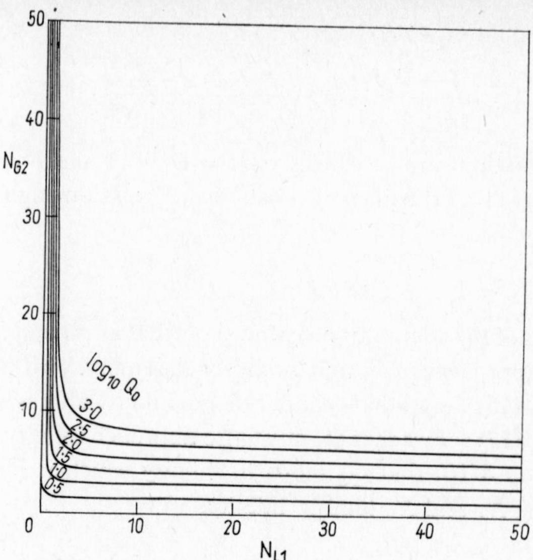

FIG. A4j.1. Existing differential contactor, N_{L1} vs. N_{G2}. Parameter Q_O. Constant $J_1 J_2 = 0 \cdot 1$.

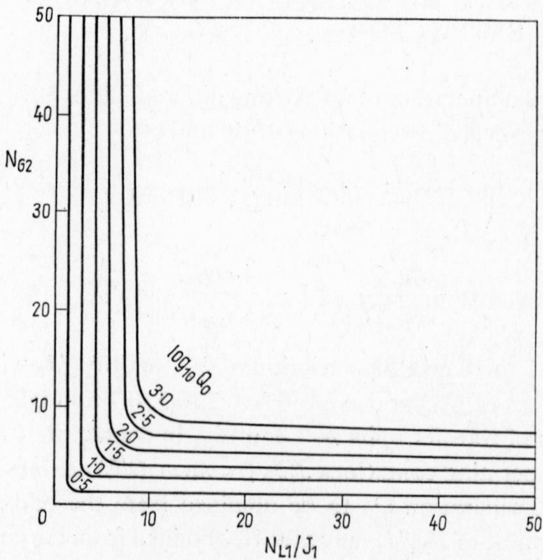

FIG. A4j.2. Existing differential contactor, N_{L1}/J_1 vs. N_{G2}. Parameter Q_O. Constant $J_1 J_2 = 0 \cdot 1$.

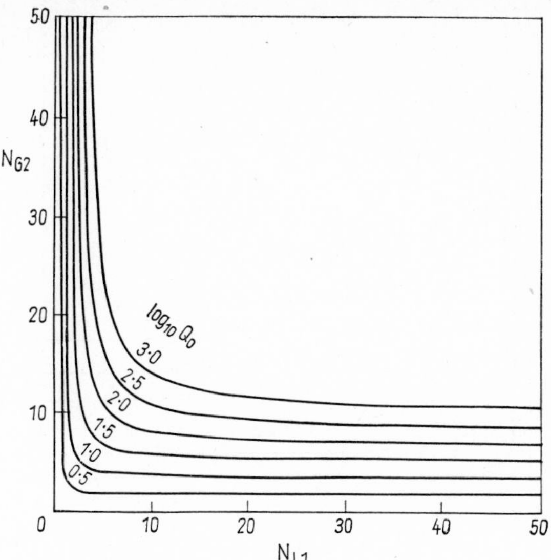

FIG. A4j.3. Existing differential contactor, N_{L1} vs. N_{G2}. Parameter Q_O. Constant $J_1J_2 = 0{\cdot}3$.

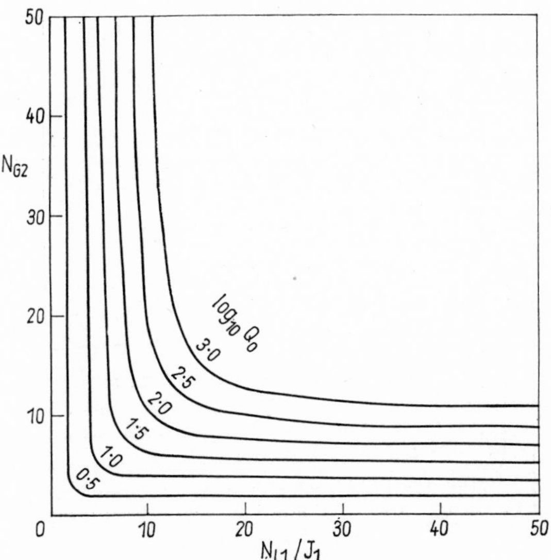

FIG. A4j.4. Existing differential contactor, N_{L1}/J_1 vs. N_{G2}. Parameter Q_O. Constant $J_1J_2 = 0{\cdot}3$.

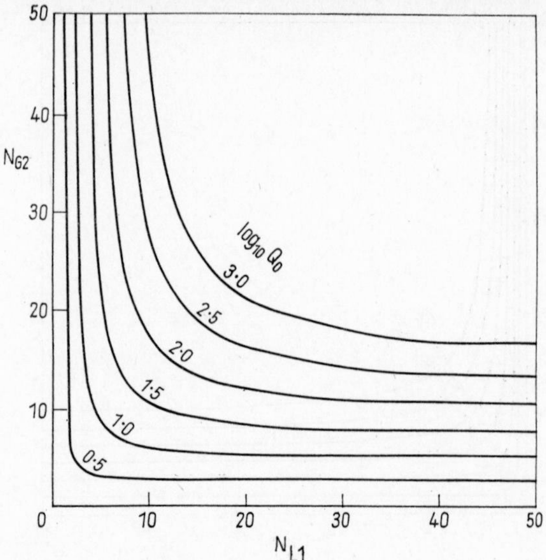

FIG. A4j.5. Existing differential contactor, N_{L1} vs. N_{G2}. Parameter Q_O. Constant $J_1 J_2 = 0{\cdot}5$.

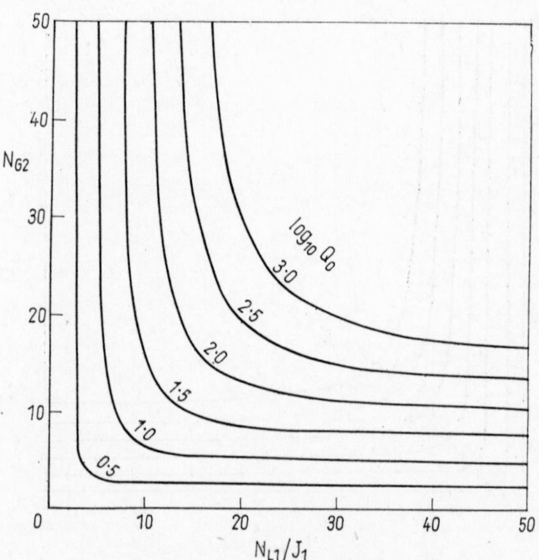

FIG. A4j.6. Existing differential contactor, N_{L1}/J_1 vs. N_{G2}. Parameter Q_O. Constant $J_1 J_2 = 0{\cdot}5$.

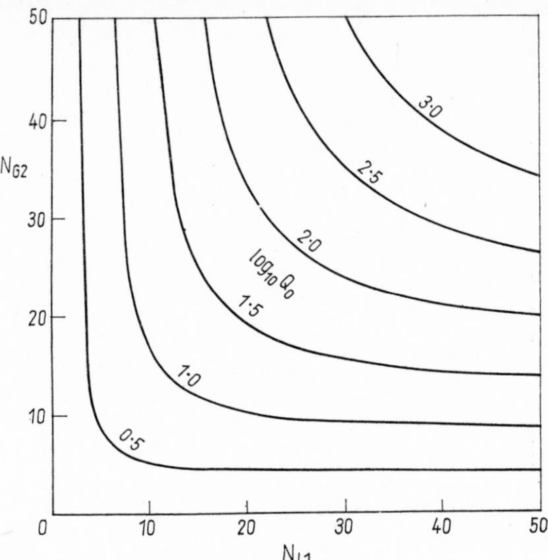

FIG. A4j.7. Existing differential contactor, N_{L1} vs. N_{G2}. Parameter Q_O. Constant $J_1J_2 = 0{\cdot}7$.

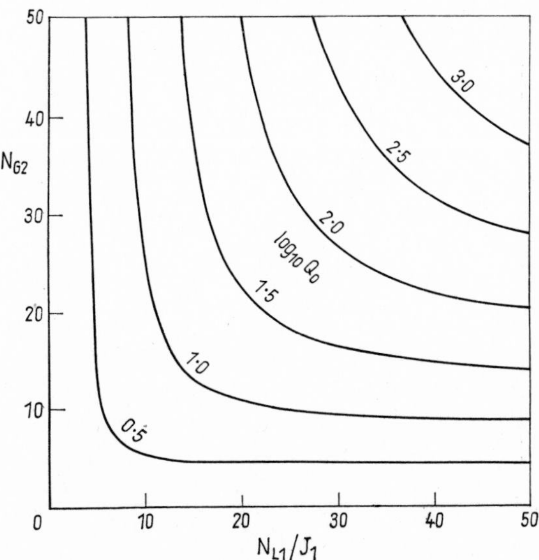

FIG. A4j.8. Existing differential contactor, N_{L1}/J_1 vs. N_{G2}. Parameter Q_O. Constant $J_1J_2 = 0{\cdot}7$.

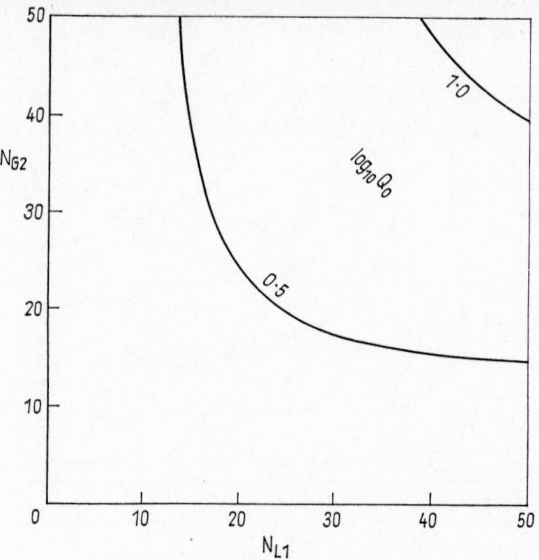

FIG. A4j.9. Existing differential contactor, N_{L1} vs. N_{G2}. Parameter Q_O. Constant $J_1J_2 = 0{\cdot}9$.

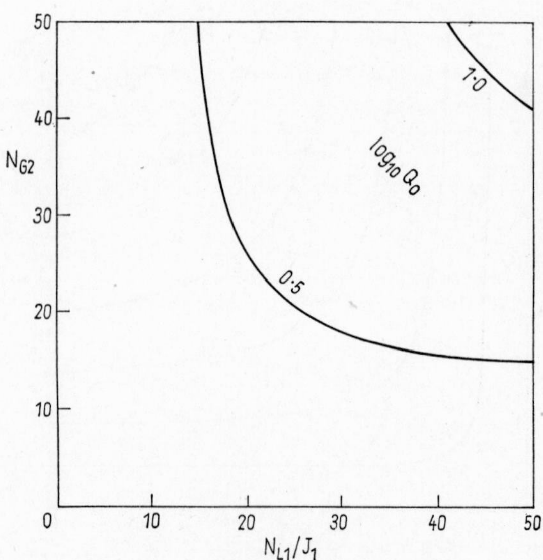

FIG. A4j.10. Existing differential contactor, N_{L1}/J_1 vs. N_{G2}. Parameter Q_O. Constant $J_1J_2 = 0{\cdot}9$.

APPENDIX 4k
SOLUTION OF THE EQUATIONS GOVERNING THE OPTIMUM OPERATING CONDITIONS IN AN EXISTING FORWARD AND BACK STAGEWISE EXTRACTOR

The optimum operation of an existing forward and back stagewise extractor is governed by equations 4(2), (6), (10), (12) and (17) which are listed respectively below,

$$N_{S1} \ln 1/J_1 = \ln Q_1 \qquad \text{A4k(1)}$$

$$N_{S2} \ln 1/J_2 = \ln Q_2 \qquad \text{A4k(2)}$$

$$\frac{1 - J_1 J_2}{Q_O - 1} = \frac{1 - J_2}{Q_2 - 1} + \frac{J_2 - J_1 J_2}{Q_1 - 1} \qquad \text{A4k(3)}$$

$$J_1 J_2 = K \qquad \text{A4k(4)}$$

$$\frac{Q_1 - 1 - N_1 (1 - J_1) Q_1}{(Q_1 - 1)^2} = \frac{Q_2 - 1 - N_2 (1 - J_2) Q_2/J_2}{(Q_2 - 1)^2} \qquad \text{A4k(5)}$$

There are thus five equations and eight variables, J_1, J_2, Q_1, Q_2, Q_O, N_1, N_2 and $J_1 J_2$. The problem is to find the value S (i.e. J_1 or J_2) which gives the maximum value of Q_O for known values of N_1 and N_2 and given operating conditions L, G, m_1 and m_2 (i.e. $J_1 J_2$).

However, mathematically speaking it is sufficient to fix any three variables in order to obtain the other five. The number of ways of choosing three from eight is

$$^3C_8 = \frac{8!}{5!\,3!} = 56$$

so there are this many different methods of solution. The one chosen will depend on its simplicity and the final form of graphical representation. Because we require to find the values of Q_O and J_1 from given values of N_1, N_2 and $J_1 J_2$ there are five variables of interest and the solution is difficult to present concisely. The form adopted is to plot N_1 and N_1/J_1 against N_2 for different values of Q_O at constant values of $J_1 J_2$. It follows that $J_1 J_2$ and Q_O must be fixed in the solution and it is convenient also to hold J_1 constant for this immediately fixes J_2. The mathematical problem is thus to obtain N_1, N_2, Q_1, Q_2 and J_2 knowing J_1J_2, Q_O and J_1.

Suppose we guess Q_2, then Q_1 will follow from equation A4k(3) and N_1 and N_2 from equations A4k(1) and (2). The value of Q_2 may thus be refined using equation A4k(5). Let us define,

$$B_1 = \frac{Q_1 - 1 - N_1(1 - J_1)Q_1}{(Q_1 - 1)^2} \qquad \text{A4k(6)}$$

$$B_2 = \frac{Q_2 - 1 - N_2(1 - J_2)Q_2/J_2}{(Q_2 - 1)^2} \qquad \text{A4k(7)}$$

and

$$g = B_1 - B_2 \qquad \text{A4k(8)}$$

Then using the perturbation procedure due to Newton[47] a better estimate of Q_2 (denoted by Q'_2) is given by

$$Q'_2 = Q_2 - g \Big/ \frac{\partial g}{\partial Q_2} \qquad \text{A4k(9)}$$

where $\partial g/\partial Q_2$ is at constant J_1, J_2 and Q_O, and is evaluated together with g at the first estimate Q_2. The second estimate Q'_2 may be refined in similar fashion and the iteration repeated until two successive estimates of Q_2 agree to within some specified limit, say 0·1%. However, first it is necessary to determine $\partial g/\partial Q_2$ at constant J_1, J_2 and Q_O. Differentiating equation A4k(8) with respect to Q_2 at constant J_1 and J_2 using equations A4k(6) and (7) gives,

$$\frac{\partial g}{\partial Q_2} = \left(\frac{2N_1Q_1(1 - J_1)}{(Q_1 - 1)^3} - \frac{1}{(Q_1 - 1)^2} - \frac{N_1(1 - J_1)}{(Q_1 - 1)^2}\right)\frac{\partial Q_1}{\partial Q_2}$$
$$- \frac{Q_1(1 - J_1)}{(Q_1 - 1)^2}\frac{\partial N_1}{\partial Q_2} + \frac{Q_2(1 - J_2)/J_2}{(Q_2 - 1)^2}\frac{\partial N_2}{\partial Q_2}$$
$$+ \frac{2N_2Q_2(1 - J_2)/J_2}{(Q_2 - 1)^3} + \frac{1}{(Q_2 - 1)^2} + \frac{N_2(1 - J_2)/J_2}{(Q_2 - 1)^2} \qquad \text{A4k(10)}$$

Now

$$\frac{\partial N_1}{\partial Q_2} = \frac{\partial N_1}{\partial Q_1}\frac{\partial Q_1}{\partial Q_2} \qquad \text{A4k(11)}$$

and it follows from equations A 4k(1) and (2) at constant J_1 and J_2 that

$$\frac{\partial N_1}{\partial Q_1} = \frac{1}{Q_1 \ln 1/J_1} \qquad \text{A 4k(12)}$$

and

$$\frac{\partial N_2}{\partial Q_2} = \frac{1}{Q_2 \ln 1/J_2} \qquad \text{A 4k(13)}$$

Differentiating equation A 4k(3) at constant J_1, J_2 and Q_O gives

$$\frac{\partial Q_1}{\partial Q_2} = \frac{(Q_1 - 1)^2}{(Q_2 - 1)} \frac{1 - J_2}{J_2 (1 - J_1)} \qquad \text{A 4k(14)}$$

Substituting equations A 4k(11), (12), (13) and (14) into equation A4k(10) leads to

$$\frac{\partial g}{\partial Q_2} = \frac{1 - J_2}{J_2 (Q_2 - 1)^2} \times$$

$$\times \left(\frac{J_2}{1 - J_2} - \frac{Q_2^2 + 1}{Q_2 - 1} \frac{1}{\ln 1/J_2} + \frac{1}{1 - J_1} - \frac{Q_1^2 + 1}{Q_1 - 1} \frac{1}{\ln 1/J_1} \right) \qquad \text{A 4k(15)}$$

which may be substituted in equation A 4k(9).

The technique described above was used to solve equations A 4k(1), (2), (3), (4) and (5) on a KDF 9 digital computer using Atlas autocode. A typical program A 14/10 with part of the output is given in Appendix 4o to illustrate the method. The program is intended to be self-explanatory rather than efficient. Graphical procedures are included and plots of N_{S1} versus N_{S2} and N_{S1}/J_1 versus N_{S2} for different values of Q_O at constant $J_1 J_2$ are given in Appendix 4l. Fortunately the method is not very sensitive to the initial estimate of Q_2.

As an alternative, the final estimate of Q_2 for one value of J_2 may be used as the initial estimate for the next successive value of J_2, returning to the original estimate at the beginning of each J_2 cycle and when Q_2 does not converge within a specified number of iterations (say 100).

It is also possible to guess N_1 or N_2 rather than Q_1 or Q_2 and to refine the initial value by a similar method to that described above for Q_2. However, this alternative is not so stable because $Q_1 = (1/J_1)^{N_1}$ and $Q_2 = (1/J_2)^{N_2}$, with the result that exponential overflows may occur during the iteration when N_1 or N_2 is estimated.

Another way of obtaining the value of Q_2 which satisfies equations A4k(1), (2), (3), (4), and (5) when J_1, J_2 and Q_O are fixed is to carry out a search over a wide range of values of Q_2. For each value of Q_2, the value of Q_1 may be obtained from equation A4k(3) and the values of N_1 and N_2 from equations A4k(1) and (2). These enable the values of B_1 and B_2 to be obtained from equations A4k(6) and (7) and hence the value of g from equation A4k(8). When g changes sign between successive values of Q_2 these must straddle the exact value of Q_2. It is thus possible to print out the values of Q_2 and the other variables when g is close to zero. The technique is similar to the graphical method of solving an equation. A typical program G14/14 written in Atlas autocode for a KDF9 digital computer is given in Appendix 4p to illustrate the method; part of the output is also shown. Once again graphical procedures are included and the program is intended to be self-explanatory rather than efficient. The search technique is not as sophisticated or rigorous as the iterative method previously described but it can be far more productive.

APPENDIX 4l
THE OPTIMUM OPERATING CONDITIONS IN AN EXISTING STAGEWISE FORWARD AND BACK EXTRACTOR

Figures A4l.1 to 10 give the variation of N_{S1} and N_{S1}/J_1 with N_{S2} at different values of Q_O for J_1J_2 = 0·1, 0·3, 0·5, 0·7 and 0·9. These were obtained as explained in Appendix 4k. Knowing the numbers of stages N_{S1} and N_{S2} in contactors 1 and 2 and the overall operating conditions J_1J_2 (= m_1m_2G/L) enables the maximum value of Q_O to be obtained from the odd-numbered figures. The value of N_{S1}/J_1 may then be obtained from the corresponding even-numbered figure, which gives J_1 (= m_1S/L) and hence the optimum value of the recirculation rate for the intermediate solvent S.

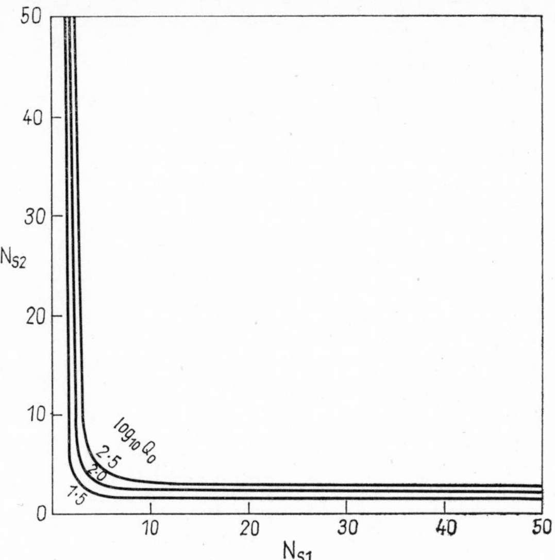

FIG. A41.1. Existing stagewise contactor, N_{S1} vs. N_{S2}. Parameter Q_O. Constant $J_1J_2 = 0{\cdot}1$.

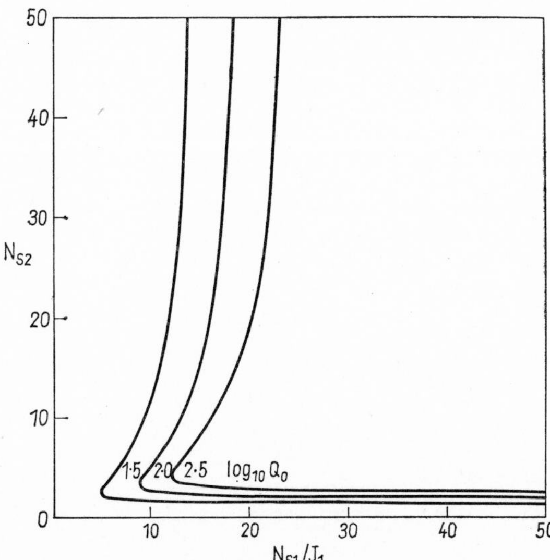

FIG. A41.2. Existing stagewise contactor, N_{S1}/J_1 vs. N_{S2}. Parameter Q_O. Constant $J_1J_2 = 0{\cdot}1$.

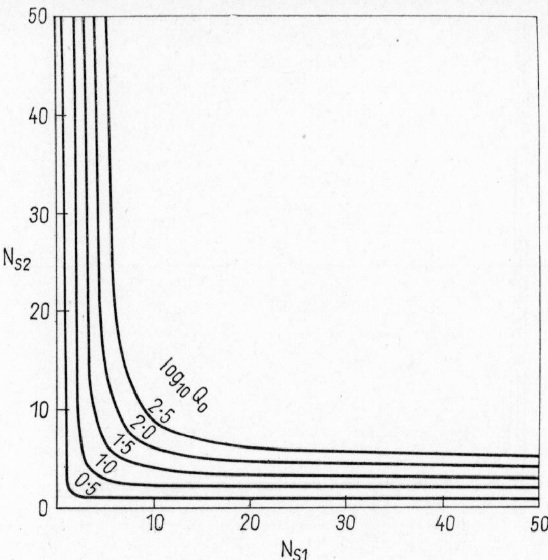

FIG. A41.3. Existing stagewise contactor, N_{S1} vs. N_{S2}. Parameter Q_O. Constant $J_1J_2 = 0{\cdot}3$.

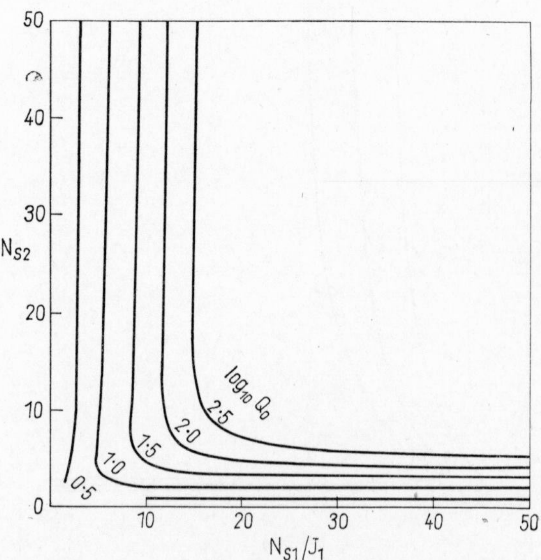

FIG. A41.4. Existing stagewise contactor, N_{S1}/J_1 vs. N_{S2}. Parameter Q_O. Constant $J_1J_2 = 0{\cdot}3$.

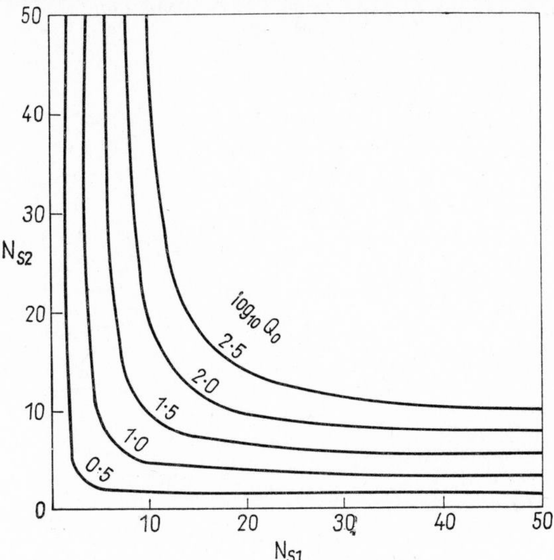

FIG. A41.5. Existing stagewise contactor, N_{S1} vs. N_{S2}. Parameter Q_O. Constant $J_1J_2 = 0{\cdot}5$.

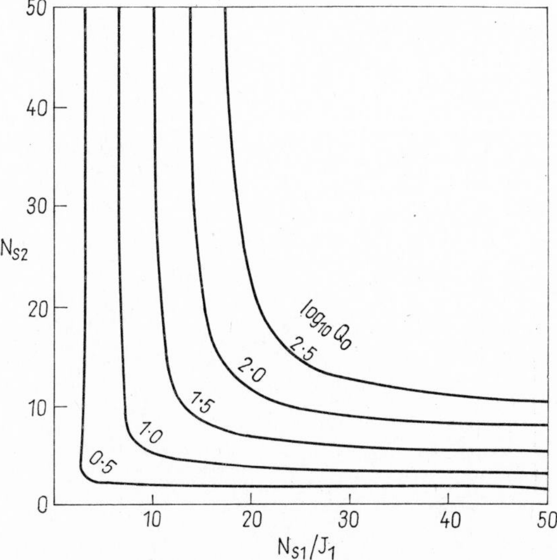

FIG. A41.6. Existing stagewise contactor, N_{S1}/J_1 vs. N_{S2}. Parameter Q_O. Constant $J_1J_2 = 0{\cdot}5$.

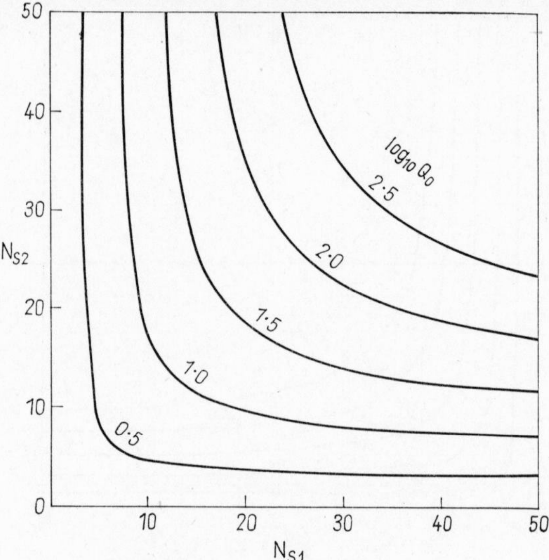

FIG. A41.7. Existing stagewise contactor, N_{S1} vs. N_{S2}. Parameter Q_O. Constant $J_1J_2 = 0·7$.

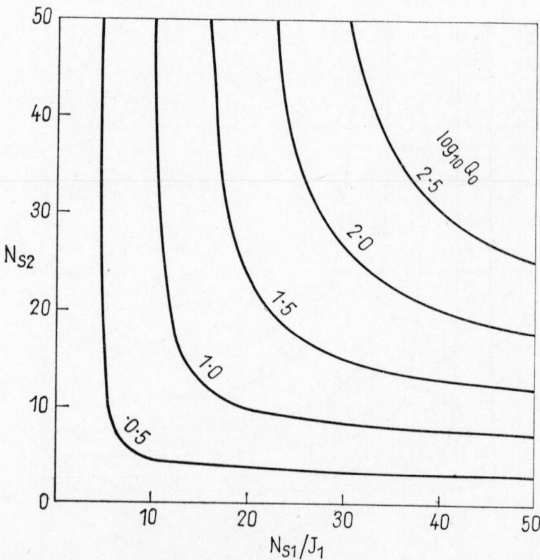

FIG. A41.8. Existing stagewise contactor, N_{S1}/J_1 vs. N_{S2}. Parameter Q_O. Constant $J_1J_2 = 0·7$.

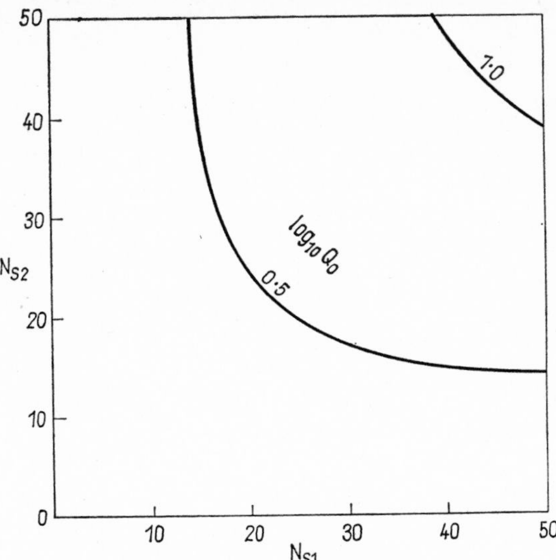

FIG. A41.9. Existing stagewise contactor, N_{S1} vs. N_{S2}. Parameter Q_O. Constant $J_1J_2 = 0{\cdot}9$.

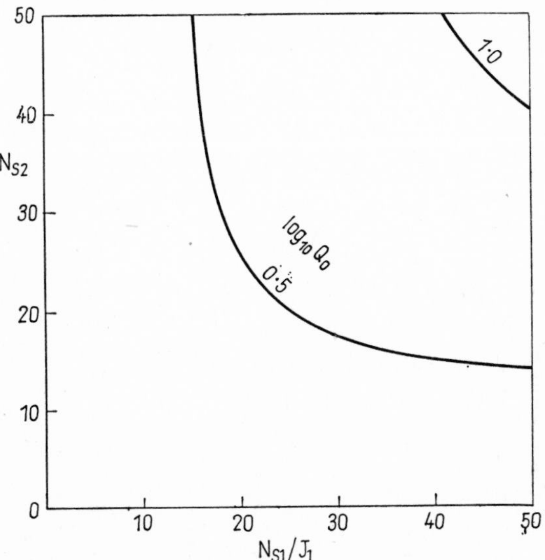

FIG. A41.10. Existing stagewise contactor, N_{S1}/J_1 vs. N_{S2}. Parameter Q_O. Constant $J_1J_2 = 0{\cdot}9$.

APPENDIX 4m
COMPUTER PROGRAM 4.1. NEWTON–RAPHSON TECHNIQUE FOR THE OPTIMUM DESIGN OF A STAGEWISE EXTRACTOR

A typical digital computer program Q 16/15, written in Atlas Autocode, is presented below, followed by a sample of the output data obtained from a KDF 9 computer. Graphical procedures are included and the program is intended to be self-explanatory rather than efficient. It includes the variables N_R, Q_R and β which relate to cross-current extraction with recirculation discussed in Chapter 5.

```
| INSERT GRAPH PLOTTER PACK; ***A

JOB
UNOTT.CHE/HARTLAND Q16/15
EXECUTION 2 MINUTES
OUTPUT O LINEPRINTER 1000 LINES
STORE 30/30 BLOCKS
COMPILER AA

begin; opengraph
caption Minimum$Number$of$Stages$Iterate$Q1$Q2ɳɳ
integer i,j,k,l,m,n
real JJ12,J2,J1,N2,Q2,A2,b,B2,N1,A1,f,fQ1,fQ2,B1,C1,C2,h,hQ1,hQ2,eQ1,eQ2,Q,B,NR,QR,Q1,S,X,Y
caption ɳ$JJ12$$$$J1$$$$J2$$$$N1$$$$N2$$$$Q$$$$B$$$$NR$$$$QR$$$$2NR/(N1+N2)$$$$Q/Qr$$$$m$$$$
caption Q1$$$$Q2ɳ; n=0
set area(1,0,10.5,10.5); draw boundary
set axes(0,0,1,5,0); move to(1,45); draw data
cycle i=7,2,9; JJ12=0.1i; newlines(4); Q1=2; Q2=20; S=2; raisepen
cycle j=100,1,1000; J1=0.00101j; J2=JJ12/J1; m=0
```

```
3: m=m+1; N1=log(Q1)/log(1/J1); N2=log(Q2)/log(1/J2)
A1=(1-J1)*Q1*log(1/J1)/(Q1-1)^2; A2=(1-J2)*Q2*log(1/J2)/(J2*(Q2-1)^2); f=A1-A2
fQ1=-(1-J1)*log(1/J1)*(Q1+1)/((Q1-1)*3); fQ2=+(1-J2)*log(1/J2)*(Q2+1)/(J2*(Q2-1)*3)
B1=(Q1-1-N1*(1-J1)*Q1)/(Q1-1)^2; B2=(Q2-1-N2*(1-J2)*Q2/J2)/(Q2-1)^2; C1=B1/A1; C2=B2/A2; h=C1-C2
hQ1=(1/(Q1*(1-J1))-1/log(1/J1))/(Q1*log(1/J1)); hQ2=(-J2/(Q2*(1-J2))+1/log(1/J2))/(Q2*log(1/J2))
eQ1=(-f*hQ2+h*fQ2)/(fQ1*hQ2-fQ2*hQ1); eQ2=(-h*fQ1+f*hQ1)/(fQ1*hQ2-fQ2*hQ1)
if mod(eQ1/Q1)<.001 and mod(eQ2/Q2)<.001 then -> 5; Q1=Q1+eQ1/S; Q2=Q2+eQ2/S
if Q1≦0 or Q2≦0 or mod(Q1-1)<1α-3 or mod(Q2-1)<1α-3 or Q1>1α10 or Q2>1α10 or N1<0 or N2<0 then -> 7
if m<100 then -> 3; caption no$convergence; newline; -> 7

5: Q=1+(1-J1J2)/((1-J2)/(Q2-1)+(J2-J1J2)/(Q1-1)); if Q≤0 then -> 6
B=J2*(1+J1)/(1+J2); NR=log(Q)/log(1/B); QR=exp((N1+N2)/2*log(1/B))
print(JJ12,2,2); print(J1,2,3); print (J2,2,3); print (N1,2,2); print(N2,2,2); print(Q,3,3)
print(B,2,3); print(NR,2,2); print(QR,3,3); print(2NR/(N1+N2),1,3); print(Q/QR,1,3)
print(m,2,0); print(Q1,3,3); print(Q2,3,3); newline; X=log(Q); Y=(N1+N2)/2
if X>10 or Y>50 or X<0 or Y<0 then -> 6; move to(X,Y); lowerpen

6: repeat; repeat; n=6

7: Q1=2; Q2=20;  -> 6 unless n=6
closegraph
endofprogram

[Q16/15]

***Z
```

Output Data

```
26/07/67   15.58.16
EDINBURGH UNIVERSITY ATLAS AUTOCODE 12/05/66

UNOTT·CHE/HARTLAND Q16/15
 0 BEGIN
28 END OF PROGRAM

PROGRAM (+PERM) OCCUPIES 5176 WORDS
PROGRAM DUMPED
COMPILING TIME 1 MIN 49 SEC/1 MIN 18 SEC
MINIMUM NUMBER OF STAGES ITERATE Q1 Q2
```

JJ12	J1	J2	N1	N2	Q	B	NR	QR	2NR/(N1+N2)	Q/QR	M	Q1	Q2
0.70	0.841	0.832	26.79	26.79	119.576	0.836	26.75	120.370	0.999	0.993	45	102.329	137.886
0.70	0.842	0.831	23.71	23.71	69.056	0.836	23.67	69.549	0.998	0.993	11	54.403	80.494
0.70	0.843	0.830	21.47	21.47	46.396	0.836	21.43	46.750	0.998	0.992	10	38.806	54.632
0.70	0.844	0.829	19.76	19.76	34.218	0.836	19.72	34.495	0.998	0.992	10	23.322	40.681
0.70	0.845	0.828	18.40	18.40	26.861	0.836	18.36	27.089	0.997	0.992	10	22.011	32.227
0.70	0.846	0.827	17.29	17.29	22.040	0.836	17.24	22.237	0.997	0.991	9	17.890	26.674
0.70	0.847	0.826	16.36	16.36	18.665	0.836	16.31	18.839	0.997	0.991	9	16.013	22.779
0.70	0.848	0.825	15.56	15.56	16.197	0.836	15.51	16.354	0.997	0.990	9	12.915	19.926
0.70	0.849	0.824	14.87	14.87	14.325	0.836	14.82	14.469	0.996	0.990	9	11.326	17.760
0.70	0.850	0.823	14.27	14.27	12.863	0.835	14.21	12.997	0.996	0.990	9	10.088	16.066
0.70	0.851	0.822	13.73	13.73	11.698	0.835	13.67	11.824	0.996	0.989	8	9.102	14.717

0.70	0.852	0.821	13.25	13.25	10.742	0.835	13.19	10.861	0.995	0.989	8	8.294	13.608
0.70	0.853	0.820	12.82	12.82	9.947	0.835	12.76	10.061	0.995	0.989	8	7.624	12.687
0.70	0.854	0.819	12.43	12.43	9.278	0.835	12.36	9.387	0.995	0.988	8	7.060	11.912
0.70	0.855	0.818	12.07	12.07	8.707	0.835	12.00	8.811	0.995	0.988	8	6.579	11.251
0.70	0.856	0.817	11.74	11.74	8.214	0.835	11.67	8.315	0.994	0.988	8	6.164	10.681
0.70	0.857	0.816	11.44	11.44	7.785	0.835	11.37	7.883	0.994	0.988	8	5.804	10.185
0.70	0.858	0.815	11.16	11.16	7.408	0.835	11.09	7.503	0.994	0.987	8	5.487	9.750
0.70	0.860	0.814	10.90	10.90	7.078	0.835	10.83	7.171	0.993	0.987	7	5.209	9.370
0.70	0.861	0.813	10.66	10.66	6.780	0.835	10.58	6.871	0.993	0.987	7	4.959	9.027
0.70	0.862	0.813	10.43	10.43	6.514	0.834	10.36	6.603	0.993	0.986	7	4.734	8.721
0.70	0.863	0.812	10.22	10.22	6.273	0.834	10.14	6.361	0.993	0.986	7	4.532	8.445
0.70	0.864	0.811	10.02	10.02	6.055	0.834	9.94	6.141	0.992	0.986	7	4.349	8.196
0.70	0.865	0.810	9.83	9.83	5.857	0.834	9.75	5.942	0.992	0.986	7	4.182	7.970
0.70	0.866	0.809	9.65	9.65	5.676	0.834	9.57	5.759	0.992	0.986	7	4.030	7.764
0.70	0.867	0.808	9.49	9.49	5.509	0.834	9.40	5.592	0.991	0.985	7	3.890	7.575
0.70	0.868	0.807	9.33	9.33	5.356	0.834	9.24	5.438	0.991	0.985	7	3.761	7.401
0.70	0.869	0.806	9.18	9.17	5.215	0.834	9.09	5.295	0.991	0.985	7	3.642	7.242
0.70	0.870	0.805	9.03	9.03	5.084	0.834	8.95	5.163	0.991	0.985	7	3.532	7.094
0.70	0.871	0.804	8.89	8.89	4.962	0.834	8.81	5.041	0.990	0.984	7	3.429	6.958
0.70	0.872	0.803	8.76	8.76	4.849	0.834	8.68	4.927	0.990	0.984	7	3.333	6.831
0.70	0.873	0.802	8.64	8.64	4.745	0.834	8.55	4.822	0.990	0.984	6	3.246	6.716
0.70	0.874	0.801	8.52	8.52	4.646	0.833	8.43	4.722	0.989	0.984	6	3.162	6.605
0.70	0.875	0.800	8.41	8.40	4.552	0.833	8.31	4.629	0.989	0.984	6	3.083	6.502
0.70	0.876	0.799	8.30	8.29	4.465	0.833	8.20	4.541	0.989	0.983	6	3.009	6.406
0.70	0.877	0.798	8.19	8.19	4.382	0.833	8.10	4.458	0.989	0.983	6	2.939	6.315
0.70	0.878	0.798	8.09	8.09	4.304	0.833	7.99	4.379	0.988	0.983	6	2.873	6.230
0.70	0.879	0.797	7.99	7.99	4.231	0.833	7.90	4.305	0.988	0.983	6	2.811	6.150
0.70	0.880	0.796	7.90	7.90	4.161	0.833	7.80	4.235	0.988	0.983	6	2.752	6.075
0.70	0.881	0.795	7.81	7.81	4.095	0.833	7.71	4.169	0.988	0.982	6	2.695	6.004

APPENDIX 4n
COMPUTER PROGRAM 4.2. RANDOM SEARCH TECHNIQUE FOR THE OPTIMUM DESIGN OF A STAGEWISE EXTRACTOR

A typical digital computer program 15/9, written in Atlas Autocode, is given opposite, followed by a sample of the output data obtained from a KDF 9 computer. The program is intended to be self-explanatory rather than efficient; it includes the variables N_R, Q_R and β which relate to cross-current extraction with recirculation discussed in Chapter 5.

```
***A

JOB
UNOTT.CHE/HARTLAND 15/9
EXECUTION 2 MINUTES
OUTPUT O LINEPRINTER 900 LINES
COMPILER AA

begin
integer i,j,k,l,n
real J2,J1,JJ12,Q,Q1,Q2,N,N1,N2,g,g',Q11,Q12,R1,R2
real A,QNN,S11,S12,a,b,c,B,QR,NR,B1,B2
caption Minimum$Number$of$Stages$–$Random$Searchɳɳ
caption ɳ$J1$$$$J2$$$$JJ12$$$$Q$$$$Q1$$$$Q2$$$$N1$$$$N2$$$$NR$$$$
caption NR/NS$$$$Q/QR$$$$B1$$$$B2$$$$g$$$$g'$$$$N1/J1ɳɳ
newlines(2); g'=1
cycle i=1,2,11; JJ12=0.1i; newlines(2)
cycle j=1,1,10; Q=exp(0.5j); print symbol(14); n=0; newline
cycle k=100,1,1000; J1=0.00101k; J2=JJ12/J1; n=n+1
R1=log(1/J1)/(1-J1); R2=log(1/J2)/(1-J2);
a=(1-J1J2)/(1-J1)*R2/R1*((1-J2)/(1-J1)*Q/J2+1)
b=R2/R1*((1-J2)*Q/(1-J1)+J2+(1-J1J2)/(1-J1))+Q-1
c=J2*R2/R1-1; if b*b-4*a*c>0 then ->4; ->1
4: S11=(b+sqrt(b*b-4*a*c))/(2*a); S12=(b-sqrt(b*b-4*a*c))/(2*a);
Q11=S11*(Q-1)+1; Q12=S12*(Q-1)+1; Q1=Q11
3: Q2=1+(1-J2)/((1-JJ12)/(Q-1)-(J2-JJ12)/(Q1-1))
if Q1>0 and Q2>0 and Q>0 then ->0; ->1
0: N1=log(Q1)/log(1/J1); N2=log(Q2)/log(1/J2)
B=J2*(1+J1)/(1+J2); N=log(Q)/log(1/B)
B1=(N1*(1-J1)*Q1+1-Q1)/(Q1-1)²
B2=(N2*(1-J2)*Q2/J2+1-Q2)/(Q2-1)²; g=B1-B2
if g/g'<0 and N1>0 and N2>0 and n>1 then ->2; ->1
2: NR=(N1+N2)/2; QR=exp(NR*log(1/B))
print(J1,2,2); print(J2,2,2); print(JJ12,2,2); print(Q,3,2) print(Q1,3,2);
print(Q2,11); print(N1,3,2); print(N2,3,2)
print(N,3,2); print(2*N/(N1+N2),2,2); print(Q/QR,1,3);
print(B1,3,3); print(B2,3,3); print(g,3,2); print(g',3,2);
print(N1/J1,3,2); newline
->1if Q1-Q12 <1α-4; Q1=Q12; ->3
1: g'=g unless g=0
repeat; repeat; repeat
end of program

***Z
```

Output Data

```
09/08/67   16.33.53
EDINBURGH UNIVERSITY ATLAS AUTOCODE   12/05/66

UNOTT.CHE/HARTLAND 15/9
   0 BEGIN
  33 END OF PROGRAM

PROGRAM (+PERM) OCCUPIES 2726 WORDS
PROGRAM DUMPED
COMPILING TIME 16 SEC/10 SEC
MINIMUM NUMBER OF STAGES – RANDOM SEARCH
```

J1	J2	JJ12	Q	Q1	Q2	N1	N2	NR	NR/NS	Q/QR	B1	B2	g	g′	N1/J1
0.68	0.15	0.10	20.09	2.36	71.0	2.21	2.23	1.96	0.88	0.670	0.172	0.172	0.00	−0.00	3.26
0.68	0.15	0.10	20.09	2.35	71.2	2.21	2.23	1.96	0.88	0.669	0.173	0.172	0.00	−0.14	3.25
0.68	0.15	0.10	20.09	2.35	71.3	2.21	2.23	1.95	0.88	0.668	0.173	0.172	0.00	−0.14	3.25
0.68	0.15	0.10	20.09	2.34	71.5	2.21	2.23	1.95	0.88	0.667	0.173	0.172	0.00	−0.14	3.24
0.68	0.15	0.10	20.09	2.33	71.6	2.21	2.23	1.95	0.88	0.666	0.174	0.172	0.00	−0.14	3.24
0.68	0.15	0.10	20.09	2.32	71.7	2.21	2.22	1.95	0.88	0.665	0.174	0.172	0.00	−0.14	3.24

0.68	0.15	0.10	20.09	2.32	71.9	2.21	2.22	1.95	0.88	0.664	0.174	0.172	0.00	−0.14	3.23
0.68	0.15	0.10	20.09	2.31	72.0	2.21	2.22	1.95	0.88	0.663	0.175	0.172	0.00	−0.14	3.23
0.69	0.15	0.10	20.09	2.30	72.2	2.21	2.22	1.95	0.88	0.662	0.175	0.171	0.00	−0.14	3.23
0.69	0.15	0.10	20.09	2.30	72.3	2.21	2.22	1.95	0.88	0.661	0.176	0.171	0.00	−0.14	3.22
0.69	0.15	0.10	20.09	2.29	72.5	2.21	2.22	1.95	0.88	0.660	0.176	0.171	0.00	−0.14	3.22
0.69	0.15	0.10	20.09	2.28	72.6	2.21	2.22	1.95	0.88	0.659	0.176	0.171	0.01	−0.14	3.21
0.69	0.14	0.10	20.09	2.28	72.7	2.21	2.22	1.95	0.88	0.658	0.177	0.171	0.01	−0.15	3.21
0.69	0.14	0.10	20.09	2.27	72.9	2.22	2.22	1.94	0.88	0.657	0.177	0.171	0.01	−0.15	3.21
0.69	0.14	0.10	20.09	2.26	73.0	2.22	2.22	1.94	0.88	0.656	0.177	0.171	0.01	−0.15	3.20
0.69	0.14	0.10	20.09	2.26	73.2	2.22	2.22	1.94	0.88	0.655	0.178	0.171	0.01	−0.15	3.20
0.69	0.14	0.10	20.09	2.25	73.3	2.22	2.22	1.94	0.88	0.654	0.178	0.171	0.01	−0.15	3.20
0.69	0.14	0.10	20.09	2.24	73.5	2.22	2.22	1.94	0.88	0.653	0.178	0.171	0.01	−0.15	3.19
0.70	0.14	0.10	20.09	2.24	73.6	2.22	2.22	1.94	0.88	0.652	0.179	0.171	0.01	−0.15	3.19
0.70	0.14	0.10	20.09	2.23	73.7	2.22	2.22	1.94	0.87	0.651	0.179	0.171	0.01	−0.15	3.19
0.70	0.14	0.10	20.09	2.22	73.9	2.22	2.21	1.94	0.87	0.650	0.180	0.170	0.01	−0.15	3.18
0.70	0.14	0.10	20.09	2.22	74.0	2.22	2.21	1.94	0.87	0.648	0.180	0.170	0.01	−0.15	3.18
0.70	0.14	0.10	20.09	2.21	74.2	2.22	2.21	1.94	0.87	0.647	0.180	0.170	0.01	−0.15	3.17
0.70	0.14	0.10	20.09	2.20	74.3	2.22	2.21	1.94	0.87	0.646	0.181	0.170	0.01	−0.15	3.17
0.70	0.14	0.10	20.09	2.20	74.5	2.22	2.21	1.94	0.87	0.645	0.181	0.170	0.01	−0.15	3.17
0.70	0.14	0.10	20.09	2.19	74.6	2.22	2.21	1.93	0.87	0.644	0.181	0.170	0.01	−0.15	3.16
0.70	0.14	0.10	20.09	2.18	74.8	2.22	2.21	1.93	0.87	0.643	0.182	0.170	0.01	−0.15	3.16
0.70	0.14	0.10	20.09	2.18	74.9	2.23	2.21	1.93	0.87	0.642	0.182	0.170	0.01	−0.15	3.16
0.71	0.14	0.10	20.09	2.17	75.1	2.23	2.21	1.93	0.87	0.641	0.182	0.170	0.01	−0.15	3.15
0.71	0.14	0.10	20.09	2.16	75.2	2.23	2.21	1.93	0.87	0.640	0.183	0.170	0.01	−0.15	3.15
0.71	0.14	0.10	20.09	2.16	75.4	2.23	2.21	1.93	0.87	0.639	0.183	0.170	0.01	−0.15	3.15
0.71	0.14	0.10	20.09	2.15	75.5	2.23	2.21	1.93	0.87	0.639	0.184	0.169	0.01	−0.15	3.14
0.71	0.14	0.10	20.09	2.15	75.6	2.23	2.21	1.93	0.87	0.638	0.184	0.169	0.01	−0.15	3.14

APPENDIX 4o
COMPUTER PROGRAM 4.3. NEWTON'S METHOD FOR THE OPTIMUM OPERATING CONDITIONS IN AN EXISTING STAGEWISE EXTRACTOR

A typical digital computer program A14/10, written in Atlas Autocode, is given below, followed by a sample of the output data obtained from a KDF 9 computer. Graphical procedures are included and the program is intended to be self-explanatory rather than efficient.

```
| INSERT GRAPH PLOTTER PACK; ***A

JOB
UNOTT.CHE/HARTLAND A14/10
EXECUTION 2 MINUTES
OUTPUT O LINEPRINTER 900 LINES
COMPILER AA

begin; opengraph
integer i,j,k,l,n
real J2,J1,JJ12,Q,Q1,Q2,N1,N2,g,B1,B2,gQ2';
caption Existing$Stagewise$Contactor$—$Newtons$Methodⱨⱨ
caption ⱨ$J1$$$$J2$$$$JJ12$$$$Q$$$$N1$$$$N2$$$$g$$$$N1/J1$$$$B1$$$$B2$$$$n$$$$Q1$$$$Q2ⱨⱨ
newlines(2);
```

```
cycle i=1,1,1;JJ12=0.5i; newlines(2)
set area(1,0,10.5,10.5); draw boundary; set axes(0,0,10,10,0); move to(10,100); draw data
cycle j=1,1,6;Q=exp(0.5j*log(10)); raisepen; print symbol(14); newline
cycle k=1,1,1001; J2=0.0101k; J1=JJ12/J2; n=0; Q2=50
3: n=n+1; Q1=1+(J2-JJ12)/((1-JJ12)/(Q-1)-(1-J2)/(Q2-1))
if Q1>0 and Q2>0 then ->0; ->1
0: N1=log(Q1)/log(1/J1)
N2=log(Q2)/log(1/J2); if N1≤0 or N2≤0 then ->1
B1=(Q1-1-N1*(1-J1)*Q1)/(Q1-1)^2; B2=(Q2-1-N2*(1-J2)*Q2/J2)/(Q2-1)^2; g=B1-B2
gQ2'=(1-J2)/(J2*(Q2-1)^2)*(J2/(1-J2)-(Q2^2+1)/((Q2-1)*log(1/J2))+1/(1-J1)-(Q1^2+1)/((Q1-1)*log(1/J1)))
Q2=Q2-g/gQ2'; if mod(g/(Q2*gQ2'))<0.001 then ->2
if n<100 then ->3; caption No$Convergence¶; ->1
2: print(J1,1,2); print(J2,1,2); print(JJ12,1,2); print(Q,3,1)
print(N1,6,2); print(N2,2,2); print(g,6,3); print(N1/J1,6,2);
print(B1,6,3); print(B2,6,3); print(n,2,0); print(Q1,3,3); print(Q2,3,3)
if N1>105 or N2>105 then ->4; move to(N1,N2); lowerpen
4: newline
1: repeat; repeat; repeat; closegraph
end of program

[A14/10]

***Z
```

Output Data

11/09/67 12.42.44
EDINBURGH UNIVERSITY ATLAS AUTOCODE 12/05/66

UNOTT.CHE/HARTLAND A14/10
0 BEGIN
24 END OF PROGRAM

PROGRAM (+PERM) OCCUPIES 4991 WORDS
PROGRAM DUMPED
COMPILING TIME 1 MIN 53 SEC/1 MIN 13 SEC
EXISTING STAGEWISE CONTACTOR – NEWTONS METHOD

J1	J2	JJ12	Q	N1	N2	9	N1/J1	B1	B2	N	Q1	Q2
0.95	0.53	0.50	10.0	26.80	3.76	−0.010	28.15	−0.276	−0.266	75	3.735	11.235
0.93	0.54	0.50	10.0	20.14	3.93	−0.007	21.57	−0.264	−0.256	66	3.952	11.648
0.92	0.55	0.50	10.0	16.34	4.11	−0.006	17.82	−0.253	−0.247	62	4.138	12.053
0.90	0.56	0.50	10.0	13.88	4.29	−0.005	15.42	−0.243	−0.238	60	4.309	12.449
0.88	0.57	0.50	10.0	12.15	4.48	−0.004	13.74	−0.235	−0.231	59	4.472	12.837
0.87	0.58	0.50	10.0	10.86	4.68	−0.004	12.51	−0.227	−0.223	58	4.625	13.225
0.85	0.59	0.50	10.0	9.88	4.88	−0.003	11.57	−0.220	−0.216	58	4.778	13.598
0.84	0.60	0.50	10.0	9.09	5.10	−0.003	10.83	−0.213	−0.210	58	4.926	13.969
0.83	0.61	0.50	10.0	8.44	5.32	−0.003	10.23	−0.206	−0.203	58	5.070	14.337
0.81	0.62	0.50	10.0	7.91	5.55	−0.003	9.74	−0.198	−0.200	58	5.212	14.702
0.80	0.63	0.50	10.0	7.45	5.80	−0.002	9.33	−0.195	−0.192	58	5.351	15.064
0.79	0.64	0.50	10.0	7.06	6.05	−0.002	8.99	−0.189	−0.187	58	5.489	15.423
0.77	0.65	0.50	10.0	6.73	6.32	−0.002	8.69	−0.184	−0.182	58	5.625	15.780
0.76	0.66	0.50	10.0	6.43	6.61	−0.002	8.44	−0.180	−0.177	58	5.759	16.134
0.75	0.67	0.50	10.0	6.17	6.91	−0.002	8.22	−0.175	−0.173	58	5.893	16.486
0.74	0.68	0.50	10.0	5.94	7.23	−0.002	8.03	−0.171	−0.169	59	6.029	16.819

0.74	0.69	0.50	10.0	5.73	7.57	−0.002	7.87	−0.167	−0.165	59	6.160	17.166
0.73	0.70	0.50	10.0	5.54	7.93	−0.002	7.72	−0.163	−0.161	59	6.291	17.511
0.72	0.71	0.50	10.0	5.37	8.32	−0.002	7.59	−0.159	−0.157	59	6.420	17.854
0.71	0.72	0.50	10.0	5.21	8.72	−0.002	7.48	−0.155	−0.154	60	6.552	18.179
0.70	0.73	0.50	10.0	5.07	9.17	−0.002	7.37	−0.152	−0.150	60	6.680	18.518
0.69	0.74	0.50	10.0	4.94	9.64	−0.002	7.28	−0.149	−0.147	60	6.808	18.855
0.68	0.75	0.50	10.0	4.82	10.15	−0.002	7.20	−0.145	−0.144	60	6.935	19.191
0.67	0.76	0.50	10.0	4.71	10.70	−0.002	7.13	−0.142	−0.141	61	7.064	19.507
0.66	0.77	0.50	10.0	4.60	11.30	−0.001	7.06	−0.140	−0.138	61	7.190	19.839
0.65	0.78	0.50	10.0	4.50	11.95	−0.001	7.01	−0.137	−0.135	61	7.315	20.170
0.64	0.79	0.50	10.0	4.41	12.67	−0.001	6.96	−0.134	−0.133	61	7.441	20.500
0.63	0.80	0.50	10.0	4.33	13.45	−0.001	6.91	−0.131	−0.130	61	7.565	20.828
0.62	0.81	0.50	10.0	4.25	14.32	−0.001	6.87	−0.129	−0.128	62	7.691	21.136
0.61	0.82	0.50	10.0	4.18	15.28	−0.001	6.83	−0.126	−0.125	62	7.815	21.461
0.60	0.83	0.50	10.0	4.11	16.35	−0.001	6.80	−0.124	−0.123	62	7.939	21.786
0.60	0.84	0.50	10.0	4.04	17.56	−0.001	6.77	−0.122	−0.121	62	8.062	22.109
0.59	0.85	0.50	10.0	3.98	18.93	−0.001	6.75	−0.120	−0.118	62	8.185	22.431
0.58	0.86	0.50	10.0	3.92	20.48	−0.001	6.73	−0.118	−0.116	63	8.309	22.731
0.58	0.87	0.50	10.0	3.86	22.28	−0.001	6.71	−0.116	−0.114	63	8.431	23.050
0.57	0.88	0.50	10.0	3.81	42.38	−0.001	6.69	−0.114	−0.112	63	8.553	23.369
0.56	0.89	0.50	10.0	3.76	26.86	−0.001	6.68	−0.112	−0.110	63	8.675	23.686
0.56	0.90	0.50	10.0	3.71	29.83	−0.001	6.66	−0.110	−0.109	63	8.796	24.003
0.55	0.91	0.50	10.0	3.66	33.46	−0.001	6.65	−0.108	−0.107	63	8.918	24.319
0.54	0.92	0.50	10.0	3.62	37.99	−0.001	6.65	−0.106	−0.105	63	9.039	24.633
0.54	0.93	0.50	10.0	3.57	43.81	−0.001	6.64	−0.105	−0.103	64	9.160	24.924
0.53	0.94	0.50	10.0	3.53	51.57	−0.001	6.64	−0.103	−0.102	64	9.280	25.236
0.53	0.95	0.50	10.0	3.49	62.43	−0.001	6.64	−0.101	−0.100	64	9.401	25.548

STOPPED AT LINE 24
UNOTT.CHE/HARTLAND A14/10
RUNNING TIME 2 MIN 13 SEC / 1 MIN 59 SEC

APPENDIX 4p
COMPUTER PROGRAM 4.4. RANDOM SEARCH TECHNIQUE FOR THE OPTIMUM OPERATING CONDITIONS IN AN EXISTING STAGEWISE EXTRACTOR

A typical digital computer program G14/14, written in Atlas Autocode, is presented below, followed by a sample of the output data obtained from a KDF 9 computer. Graphical procedures are included and the program is intended to be self-explanatory rather than efficient.

```
1 INSERT GRAPH PLOTTER PACK; ***A

JOB
UNOTT.CHE/HARTLAND G14/14
EXECUTION 2 MINUTES
OUTPUT 0 LINEPRINTER 900 LINES
COMPILER AA

begin; opengraph
integer i,j,k,l,n
real J2,J1,JJ12,Q,Q1,Q2,N1,N2,g,g',B1,B2;
caption Existing$Stagewise$Contactor$−$Random$Searchɳɳ
caption ɳ$J1$$$$J2$$$$JJ12$$$$Q$$$$N1$$$$N2$$$$g$$$$g'$$$$
caption N1/J1$$$$Q1$$$$Q2$$$$B1$$$$B2ɳɳ
newlines(2); g'=1
cycle i=1,1,1; JJ12=0.7i; newlines(2)
set area(1,0,10.5,10.5); draw boundary
set axes(0,0,10,10,0); move to (10,100); draw data
cycle j=1,1,6; Q=exp(0.5j*log(10)); raisepen; print symbol(14)
newline
cycle k=1,2,101; J2=0.0101k; J1=JJ12/J2; n=0
cycle 1=1,10,1001; Q2=1.0011
n=n+1; Q1=1+(J2−JJ12)/((1−JJ12)/(Q−1)−(1−J2)/(Q2−1))
if Q1>0 and Q2>0 then −>0; −>1
0: N1=log(Q1)/log(1/J1)
N2=log(Q2)/log(1/J2)
B1=(Q1−1−N1*(1−J1)*Q1)/(Q1−1)²
B2=(Q2−1−N2*(1−J2)*Q2/J2)/(Q2−1)²; g=B1−B2
if N1>0 and N2>0 and g/g'<0 and n>1 then −>2;−>1
2: print(J1,1,2); print(J2,1,2); print(JJ12,1,2); print(Q,3,1); print(N1,6,2);
print(N2,2,2); print(g,6,3); print(g',6,3); print(N1/J1,6,2); print(Q1,3,3)
print(Q2,3,3); print(B1,6,3); print(B2,6,3); if N1>105 or N2>105 then −>3
move to(N1,N2); lowerpen
3: newline
1: g'=g unless g=0
repeat
repeat
repeat
repeat
closegraph
end of program
[G14/14]
***Z
```

Output Data

```
17/08/67   12.51.06
EDINBURGH UNIVERSITY ATLAS AUTOCODE   12/05/66

UNOTT.CHE/HARTLAND G14/14
     0   BEGIN
    32   END OF PROGRAM

PROGRAM (+PERM) OCCUPIES 4949 WORDS
PROGRAM DUMPED
COMPILING TIME   2 MIN 6 SEC / 1 MIN 13 SEC
EXISTING STAGEWISE CONTACTOR – RANDOM SEARCH
```

J1	J2	JJ12	Q	N1	N2	g	g′	N1/J1	Q1	Q2	B1	B2
0.98	0.72	0.70	100.0	183.96	13.88	0.004	−0.478	188.45	84.759	101.101	−0.041	−0.045
0.98	0.72	0.70	100.0	150.81	14.17	−0.030	0.004	154.50	38.087	111.111	−0.073	−0.042
0.95	0.74	0.70	100.0	111.54	14.80	0.034	−10.477	117.48	327.156	91.091	−0.014	−0.048
0.95	0.74	0.70	100.0	78.50	15.46	−0.011	0.007	82.68	58.872	111.111	−0.053	−0.041
0.92	0.76	0.70	100.0	129.32	15.83	0.051	−9.408	139.94	27135.589	81.081	−0.000	−0.052
0.92	0.76	0.70	100.0	53.90	16.96	−0.005	0.007	58.32	70.446	111.111	−0.045	−0.041
0.90	0.78	0.70	100.0	54.39	17.48	0.036	−8.394	60.43	306.468	81.081	−0.015	−0.051

0.90	0.78	0.70	100.0	41.37	18.74	−0.001	0.008	45.96	77.822	111.111	−0.041	−0.040
0.88	0.80	0.70	100.0	49.72	18.88	0.047	−7.430	56.68	671.155	71.071	−0.008	−0.055
0.88	0.80	0.70	100.0	32.84	21.25	−0.006	0.000	37.44	73.635	121.121	−0.042	−0.037
0.86	0.82	0.70	100.0	71.45	20.48	0.060	−6.513	83.51	68846.804	61.061	−0.000	−0.060
0.86	0.82	0.70	100.0	28.02	23.89	−0.003	0.002	32.75	78.903	121.121	−0.040	−0.036
0.84	0.84	0.70	100.0	33.37	23.31	0.048	−5.639	39.96	410.123	61.061	−0.011	−0.059
0.84	0.84	0.70	100.0	24.52	27.20	−0.002	0.002	29.36	83.118	121.121	−0.038	−0.036
0.82	0.86	0.70	100.0	32.63	25.78	0.060	−4.807	40.02	781.071	51.051	−0.006	−0.067
0.82	0.86	0.70	100.0	21.86	31.44	−0.001	0.003	26.80	86.568	121.121	−0.036	−0.035
0.80	0.88	0.70	100.0	53.61	28.73	0.076	−4.013	67.30	196772.487	41.041	−0.000	−0.077
0.80	0.88	0.70	100.0	19.60	37.71	−0.003	0.000	24.60	86.170	131.131	−0.036	−0.033
0.78	0.90	0.70	100.0	23.90	34.85	0.064	−3.254	30.69	394.557	41.041	−0.011	−0.075
0.78	0.90	0.70	100.0	17.96	45.75	−0.002	0.001	23.06	89.267	131.131	−0.034	−0.032
0.76	0.92	0.70	100.0	23.80	40.72	0.083	−2.529	31.25	652.269	31.031	−0.007	−0.090
0.76	0.92	0.70	100.0	16.60	57.80	−0.001	0.001	21.80	91.965	131.131	−0.033	−0.032
0.75	0.94	0.70	100.0	18.61	54.85	0.073	−1.835	24.97	238.152	31.031	−0.016	−0.089
0.75	0.94	0.70	100.0	15.46	77.87	−0.001	0.002	20.75	94.336	131.131	−0.032	−0.031
0.73	0.96	0.70	100.0	17.62	73.66	0.098	−1.171	24.15	258.587	21.021	−0.015	−0.113
0.73	0.96	0.70	100.0	14.49	117.95	−0.000	0.002	19.86	96.437	131.131	−0.031	−0.031
0.71	0.98	0.70	100.0	16.76	116.97	0.153	−0.535	23.46	279.993	11.011	−0.014	−0.166
0.71	0.98	0.70	100.0	13.64	241.35	−0.001	0.000	19.09	97.935	141.141	−0.030	−0.029
0.70	1.00	0.70	100.0	12.92	23987.75	0.134	−0.025	18.46	100.294	11.011	−0.029	−0.164
0.70	1.00	0.70	100.0	12.91	49495.12	−0.001	0.001	18.45	99.990	141.141	−0.029	−0.028

CHAPTER 5

Forward and Back Extraction Using Cross-current Flow with Recirculation

5.1. Introduction

The optimisation of forward and back extraction using counter-current flow in stagewise and differential extractors is discussed in the preceding chapter. Another way of carrying out forward and back

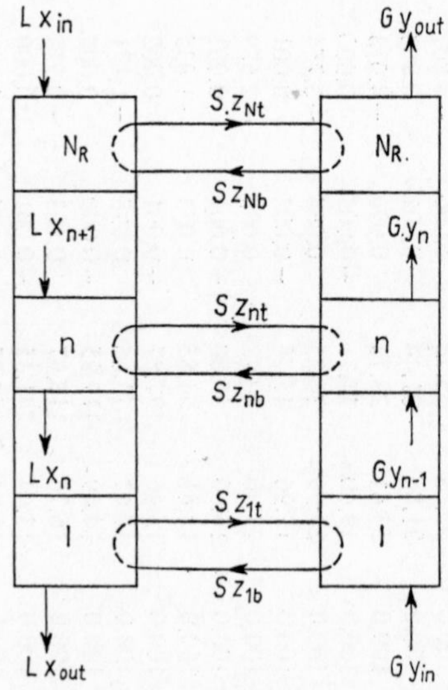

FIG. 5.1. Flows and concentrations in cross-current extraction with recirculation.

extraction in a stagewise process is by recirculating the intermediate solvent between pairs of stages in contactors 1 and 2 as shown in Fig. 5.1. This will be called cross-current extraction with recirculation.

The composition and flowrate of the intermediate solvent is under direct control in each pair of stages and can be varied along the contactor.

Such a plant has been used to investigate the possibility of extracting phenols from tar oils using aqueous sodium phenate as the intermediate solvent and benzole as the final solvent.[1] The number of stages in contactors 1 and 2 of Fig. 5.1 must obviously be equal so no optimum exists in this case. For given operating conditions and total number of stages there is a fixed separation. However, it is interesting to compare this with the corresponding maximum separation using counter-current flow. Alternatively one could compare the number of pairs of stages required to give this separation with the minimum total number using counter-current flow.

The number of pairs of theoretical cross-current extraction stages required for a given separation must first be calculated. It is assumed that the flowrate and composition of the intermediate solvent is constant for all the pairs of stages; the equilibrium data is linear in each contactor and the phases are immiscible.

5.2. Theory of Cross-current Extraction with Recirculation

Consider a cross-current extractor with N_R pairs of stages as shown in Fig. 5.1. The flowrates L, G and S and concentrations x, y and z are as for counter-current flow. The concentrations in the intermediate solvent flowing between the nth pair of stages are denoted z_{nt} and z_{nb}. These are in equilibrium with the stage that they leave. The rest of the nomenclature is identical with that for counter-current extraction.

The linear equilibrium relationships

$$z = m_1 x + c_1$$

and

$$y = m_2 z + c_2$$

are assumed in contactors 1 and 2 respectively.

Using these to eliminate z_{nt} and z_{nb} from solute balances round each of the stages in the nth pair gives

$$y_n = \frac{m_1 m_2}{J_1} [(1 + J_1) x_n - x_{n+1}] + m_2 c_1 + c_2 \qquad 5(1)$$

and

$$m_1 m_2 x_n = (1 + J_2)\, y_n - J_2 y_{n-1} - (m_2 c_1 + c_2) \qquad 5(2)$$

where $J_1 = m_1 S/L$ and $J_2 = m_2 G/S$ are the extraction factors in contactors 1 and 2 respectively. Equations 5(1) and (2) may be regarded as the equilibrium relationships for the pair of stages treated as a whole. Equation 5(1) may be used to eliminate y_n and y_{n-1} from a solute balance round the nth pair of stages to give the finite difference equation:

$$(1 + J_2)\, x_{n+1} - (1 + J_1 J_2 + 2J_2)\, x_n + J_2\, (1 + J_1)\, x_{n-1} = 0$$

The general solution[(2)] to this is

$$x_n = \mathscr{A}\beta^n + \mathscr{B}$$

where

$$\beta = \frac{J_2\,(1 + J_1)}{1 + J_2} \qquad 5(3)$$

$\mathscr{A}$ and $\mathscr{B}$ are constants which may be obtained from the boundary conditions,

$$n = N_R + 1, \quad x_{N_R+1} = x_{\text{in}}$$

$$n = 0, \quad y_0 = y_{\text{in}}$$

so that from equation 5(1)

$$y_{\text{in}} = \frac{m_1 m_2}{J_1}\,[(1 + J_1)\, x_0 - x_1] + m_2 c_1 + c_2$$

This gives for the x profile

$$\frac{x_{\text{in}} - x_n}{x_{\text{in}} - x(y_{\text{in}})} = \frac{\beta^{N_R+1} - \beta^n}{\beta^{N_R+1} - \beta/J_1 J_2} \qquad 5(4)$$

where $x(y_{\text{in}})$ is the x concentration in equilibrium with y_{in} given by

$$x(y_{\text{in}}) = y_{\text{in}}/m_1 m_2 - c_1/m_1 - c_2/m_1 m_2$$

Substituting x_n and x_{n+1} from equation 5(4) into equation 5(1) gives the y profile as

$$\frac{y_n - y_{\text{in}}}{y(x_{\text{in}}) - y_{\text{in}}} = \frac{\beta^n - 1}{J_1 J_2 \beta^{N_R} - 1} \qquad 5(5)$$

where $y(x_{\text{in}})$ is the y concentration in equilibrium with x_{in} given by

$$y(x_{\text{in}}) = m_1 m_2 x_{\text{in}} + m_2 c_1 + c_2$$

Introducing the boundary condition

$$n = 1, \quad x_1 = x_{out}$$

equation 5(4) becomes

$$N_R \ln \beta = \ln \frac{Q_{6/2}/J_1J_2 - 1}{Q_{6/2} - 1} \qquad 5(6)$$

and when $n = N_R$, $y_N = y_{out}$ so equation 5(5) becomes

$$N_R \ln 1/\beta = \frac{J_1J_2Q_{1/2} - 1}{Q_{1/2} - 1} \qquad 5(7)$$

where $Q_{1/2}$ and $Q_{6/2}$ are inverse overall separation factors defined by Table 2.3. Equations 5(6) and (7) are analogous to the simpler ones for a single counter-current extractor derived by Souders and Brown.[(3)] They were derived in this form as the inlet concentrations are more likely to be known than the outlet concentrations and so equations 5(4) and (5) are the most useful ways of expressing the concentration profiles. $Q_{1/2}$ and $Q_{6/2}$ are related by an overall balance involving J_1J_2 to the other twelve separation factors and equations 5(6) and (7) may be expressed in any of the eight basic forms listed in Table 2.3 if J is replaced by J_1J_2. In a form analogous to the original equation derived by Kremser[(4)], equation 5(7) becomes

$$N_R \ln 1/\beta = \ln [(1 - J_1J_2)\, Q_{2/4} + J_1J_2] \qquad 5(8)$$

where

$$Q_{2/4} = \frac{y_{in} - y(x_{in})}{y_{out} - y(x_{in})}$$

When $J_1J_2 = 1$, equation 5(3) shows that $\beta = 1$ and equation 5(8) reduces to

$$N_R = (1 + J_2)(Q_{2/4} - 1) \qquad 5(9)$$

Alternatively in the form recommended by Chapter 2, equations 5(6) and (7) become

$$N_R \ln 1/\beta = \ln Q_R \qquad 5(10)$$

where

$$Q_R = \frac{y_{in} - y(x_{out})}{y_{out} - y(x_{in})} \qquad 5(11)$$

is a separation factor identical with that used in counter-current extraction (with subscript R to indicate its application to cross-current extraction with recirculation).

Physical inversion of Fig. 5.1 shows that all the quantities may be inverted as for counter-current extraction. Inversion is denoted by *. The inverses of J_1 and J_2 are $J_1^* = 1/J_2$ and $J_2^* = 1/J_1$ and so the inverse of β is $\beta^* = 1/\beta$, which is evident from equations 5(6) and (7). The inverse of Q_R is $Q_R^* = 1/Q_R$ and N_R is unchanged on inversion. Thus equation 5(10) may be rewritten in the identical form

$$N_R \ln 1/\beta^* = \ln Q_R^* \qquad 5(12)$$

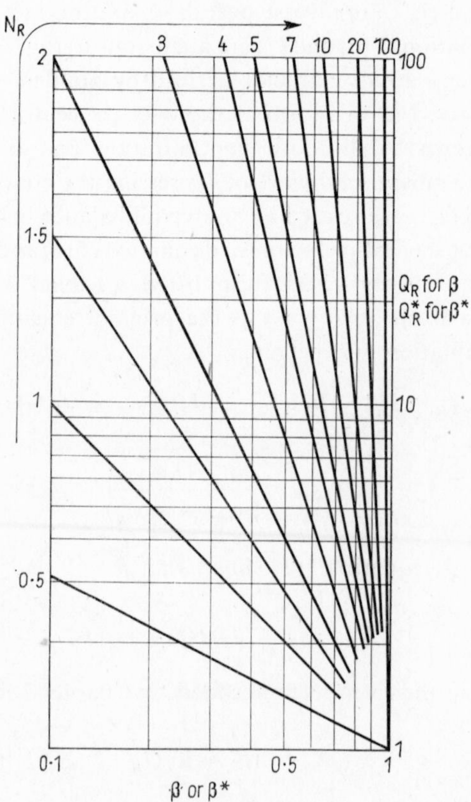

FIG. 5.2. Variation of the separation Q_R with $\beta = \dfrac{J_1 J_2 + J_2}{1 + J_2}$ for different numbers of stages N_R.

This is useful when β is greater than one for then β^* is less than one and so only the region less than one need be considered. Note that β is less than one when J_1J_2 is less than one and β is greater than one when J_1J_2 is greater than one.

Equation 5(10) is plotted in Fig. 5.2 which is identical with that for a single counter current-extractor given in Fig. 2.1 if N_S, Q and J are replaced respectively by N_R, Q_R and β. When β is greater than one, equation 5(11) indicates that $1/\beta$ and $1/Q_R$ respectively should be used instead of β and Q_R in Fig. 5.2. Plots involving J_1 and J_2 are also presented in Appendix 5a, but as there are then four variables of interest, N_R, Q_R, J_1 and J_2, they cannot be related on a single graph.

5.3. Comparison with Counter-current Forward and Back Extraction

For forward and back extraction using counter-current flow the minimum total number of stages occurs when the numbers of stages in each contactor are for practical purposes equal, so that $N_{S1} = N_{S2} = N_{SO}$. The relationship between N_{SO}, the separation Q_O and the optimum values of J_1 and J_2 is presented in Chapter 4, and Q_O is defined in identical fashion to Q_R as given by equation 5(10).

For a given separation ($Q_O = Q_R$), N_R is compared with N_{SO} in Fig. 5.3. N_R is calculated at the optimum values of J_1 and J_2 for counter-current flow. Figure 5.3 shows that when J_1J_2 is less than one, N_R is less than N_{SO} but only significantly so when J_1J_2 or Q_O is small. As J_1J_2 approaches 1 the ratio N_R/N_{SO} approaches 1.

The equations governing the optimisation of counter-current flow also give the maximum separation Q_O for a given value of N_S. Q_O is compared with Q_R when $N_R = N_{SO}$ in Fig. 5.4, Q_R being calculated at the optimum values of J_1 and J_2 for counter-current flow. The variation of Q_O/Q_R is greater than that of N_R/N_{SO} but the behaviour of $\ln Q_O/\ln Q_R$ is similar. In fact N_R/N_{SO} may be replaced by $\ln Q_O/\ln Q_R$ in Fig. 5.3 if $Q_O = Q_R$ is replaced by Q_O.

When N_{SO} is large the equations governing the optimisation of counter-current flow reduce to

$$N_{SO} \ln 1/\sqrt{(J_1J_2)} = \ln Q_O \qquad 5(13)$$

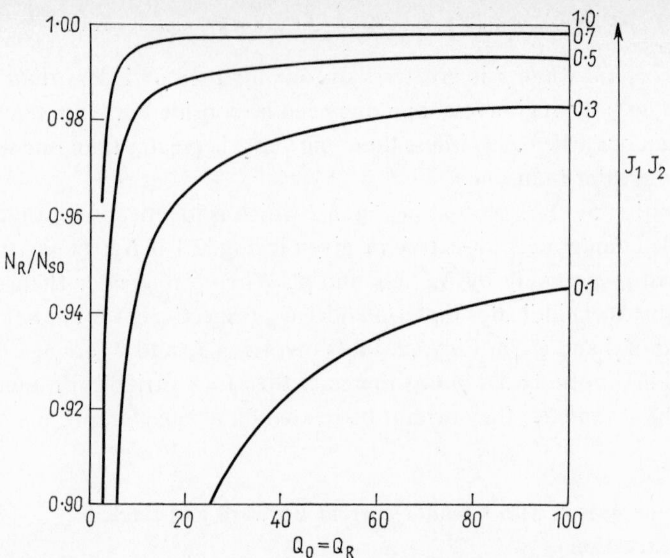

FIG. 5.3. Comparison of forward and back extraction using counter-current flow and cross flow with recirculation. Variation of ratio N_R/N_{SO} with $Q_O = Q_R$ for different values of J_1J_2.

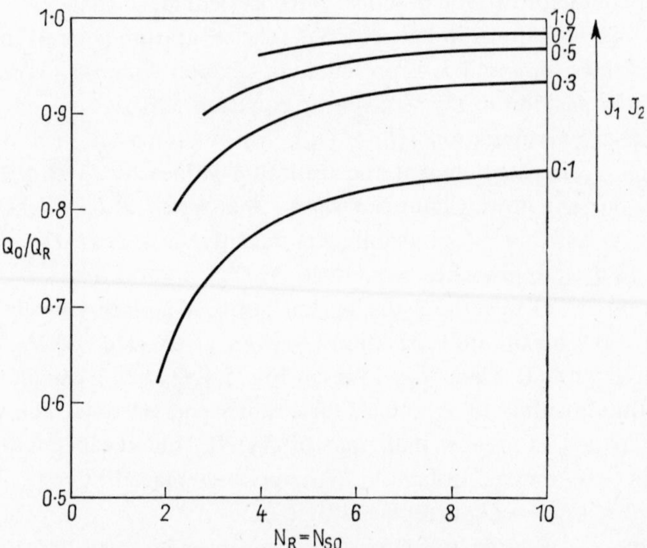

FIG. 5.4. Comparison of forward and back extraction using counter-current flow and cross flow with recirculation. Variation of ratio Q_O/Q_R with $N_R = N_{SO}$ for different values of J_1J_2.

and

$$J_1 = J_2 = \surd(J_1 J_2) \qquad 5(14)$$

Thus, for a given separation, $Q_O = Q_R$,

$$\frac{N_R}{N_{SO}} = \frac{\ln 1/\surd(J_1 J_2)}{\ln 1/\beta} \qquad 5(15)$$

and for a given number of stages, $N_R = N_{SO} = N$,

$$\frac{Q_O}{Q_R} = \left(\frac{\beta}{\surd(J_1 J_2)}\right)^N \qquad 5(16)$$

Both these ratios are unity when β is calculated at the optimum values of $J_1 = J_2$.

Bearing in mind the added complexity of cross-current extraction it seems unlikely that it will be more attractive economically than counter-current extraction when J_1 and J_2 are constant throughout the system. However, it must be remembered that one of the advantages of cross-current extraction is that the operating conditions (m_1, m_2, c_1, c_2 and S) can be directly controlled in each pair of stages, so that J_1 and J_2 need not be constant.

Unlike counter-current extraction, increasing the recirculation rate S for given overall operating condition $J_1 J_2$ always increases the separation Q_R from a given number of pairs of stages N_R. For as S increases $J_2 = m_2 G/S$ decreases, and so $\beta = (J_2 + J_1 J_2)/(J_2 + 1)$ decreases when $J_1 J_2$ is less than one, and Q_R given by equation 5(10) increases. When $J_1 J_2$ is greater than one, β increases and so $\beta^* = 1/\beta$ decreases, and Q_R given by equation 5(12) increases. Thus whilst it is instructive to compare cross-current and counter-current extraction, at the optimum value of J_1 and J_2 for counter-current extraction, these values have no particular significance in cross-current extraction.

When $J_1 = J_2 = \surd(J_1 J_2)$ the value of β reduces to $\surd(J_1 J_2)$ and so equation 5(10) becomes

$$N_R \ln 1/\surd(J_1 J_2) = \ln Q_R \qquad 5(17)$$

Comparing this with equation 5(13) shows that, when N_S is large, the ratios N_R/N_{SO} and Q_O/Q_R both equal unity at constant $Q_O = Q_R$ and $N_{SO} = N_R$ respectively.

When there is no recirculation S is zero and $J_2 = m_2G/S$ approaches infinity so that β (given by equation 5(3)) is one. For a given number of stages it follows from equation 5(10) that Q_R is one, so as expected there is no separation.

When the recirculation is infinite J_2 is zero and $\beta = J_1J_2$, which is the smallest value β can attain when J_1J_2 is less than 1 and is fixed by the overall operating conditions. The minimum number of stages N_R for a given separation Q_R or the maximum value of Q_R for a given N_R are thus related by

$$N_R \ln 1/J_1J_2 = \ln Q_R \tag{5(18)}$$

These optimum values may be compared with the values of N_{SO} and Q_O calculated at the optimum values of J_1 and J_2 for counter-current flow.

When N_{SO} is large it is related to Q_O by equation 5(13) and so for a given separation $Q_O = Q_R$,

$$\frac{N_R}{N_{SO}} = \frac{1}{2}$$

and for a given number of stages $N_{SO} = N_R = N$,

$$\frac{\ln Q_O}{\ln Q_R} = \frac{1}{2}$$

which may be combined with equation 5(18) to give,

$$\ln Q_R/Q_O = N \ln 1/\sqrt{(J_1J_2)}$$

These equations cover most of the practical range. When N_{SO} is small the ratios N_R/N_{SO} and $\ln Q_O/\ln Q_R$ lie between 0·5 and 1·0 showing that cross-current forward and back extraction is always more efficient than counter-current forward and back extraction.

5.4. Comparison with Ordinary Counter-current Extraction

It is also interesting to compare forward and back extraction using cross-current flow with ordinary counter-current extraction using no intermediate solvent. Counter-current extraction is governed by equation 2(11) which is

$$N_S \ln 1/J = \ln Q$$

where N_S is the number of stages, $J = mG/L$ and Q is identical with Q_R defined by equation 5(11). Cross-current extraction is governed by equation 5(10) which is

$$N_R \ln 1/\beta = \ln Q_R$$

where

$$\beta = \frac{J_2\,(1 + J_1)}{1 + J_2}$$

Comparing the two types of extraction under the same condition of individual extraction $J = J_1 = J_2$ shows that when the separations Q and Q_R are equal so are the number of stages N_S and N_R. (This comparison is not favourable to counter-current extraction as there are N_R stages in both the forward and back extractors.)

Alternatively comparison under the same conditions of overall extraction so that $J = J_1 J_2$ shows that for a given separation $Q = Q_R$,

$$\frac{N_S}{N_R} = \frac{\ln\,[(1 + J_2)/(J_2 + J)]}{\ln 1/J}$$

and for a given number of stages $N_S = N_R$

$$\frac{\ln Q_R}{\ln Q} = \frac{\ln\,[(1 + J_2)/(J_2 + J)]}{\ln 1/J}$$

These ratios are plotted against J for different values of the parameter J_2 in Fig. 5.5 which shows that counter-current extraction is always more efficient than forward and back extraction using cross flow but that the ratios approach unity as J_2 approaches zero. This corresponds to infinite recirculation of the intermediate solvent in cross flow which gives the maximum separation Q_R from a given number of stages N_R or the minimum N_R for a given Q_R.

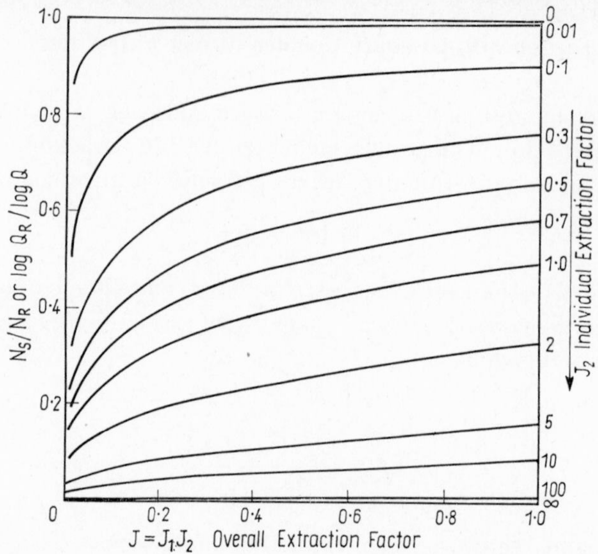

FIG. 5.5. Comparison of cross flow forward and back extraction with ordinary counter-current extraction. Variation of ratios N_S/N_R or $\ln Q_R/\ln Q$ with $J = J_1J_2$ for different values of J_2 at a given separation $Q = Q_R$ or number of stages $N_S = N_R$.

Example 5.1

In Example 4.1 of the preceding chapter on counter-current flow for a separation $Q_O = 1/83{\cdot}6$ the minimum number of stages $N_S = 4{\cdot}8$ and the optimum values of $J_1 = 2{\cdot}86$ and $J_2 = 2{\cdot}10$ so that $J_1J_2 = 6$.

For cross-current extraction

$$\beta = \frac{J_2(1 + J_1)}{1 + J_2} = \frac{2{\cdot}1 + 6}{1 + 2{\cdot}1} = 2{\cdot}61$$

or $\beta^* = 1/\beta = 0{\cdot}383$ and so the number of pairs of stages to give the separation $Q_R^* = 83{\cdot}6$ is given by Fig. 5.2 or equation 5(12) as $N_R = 4{\cdot}6$. The ratio $N_R/N_{SO} = 4{\cdot}6/4{\cdot}8 = 0{\cdot}96$ which agrees with the value indicated by Fig. 5.3 with $Q_R^* = 83{\cdot}6$ and $(J_1J_2)^* = 1/J_2J_1 = 0{\cdot}17$.

The separation from 4·8 pairs of stages with $\beta = 2{\cdot}61$ is given by Fig. 5.2 or equation 5(12) as $Q_R^* = 100$. The ratio $Q_O^*/Q_R^* = 83{\cdot}6/100 = 0{\cdot}83$ which agrees with the value indicated by Fig. 5.4 when $N_{SO} = N_R = 4{\cdot}8$ and $(J_1J_2)^* = 0{\cdot}17$.

When $J_1 = J_2$ the value of β is given by $\sqrt{(J_1 J_2)} = \sqrt{6} = 2{\cdot}45$ or $\beta^* = 0{\cdot}408$. Equation 5(17) or Fig. 5.2 indicate that when $Q_R^* = 83{\cdot}6$ the value of N_R is 4·94 and when $N_R = 4{\cdot}8$ the value of Q_R^* is 73·5. Thus at constant $Q_O^* = Q_R^* = 83{\cdot}6$ the ratio $N_R/N_{SO} = 4{\cdot}94/4{\cdot}8 = 1{\cdot}03$ and at constant $N_{SO} = N_R = 4{\cdot}8$ the ratio $Q_O^*/Q_R^* = 1{\cdot}14$. These ratios are greater than those above as the recirculation rate used above must be decreased to make J_1 and J_2 equal.

When the recirculation is infinite the value of β is given by $J_1 J_2 = 6$ or $\beta^* = 0{\cdot}167$. Equation 5(18) or Fig. 5.2 indicate that with $Q_R^* = 83{\cdot}6$ the minimum value of N_R is 2·47 and with $N_R = 4{\cdot}8$ the maximum value of Q_R^* is 5080. Thus at constant $Q_O^* = Q_R^* = 83{\cdot}6$ the ratio $N_R/N_{SO} = 2{\cdot}47/4{\cdot}8 = 0{\cdot}515$ and at constant $N_{SO} = N_R = 4{\cdot}8$ the ratio $Q_O^*/Q_R^* = 0{\cdot}0165$. These ratios indicate that under optimum conditions cross-current flow with recirculation is much more effective than counter-current flow. However, one must bear in mind that the cost of recirculation is of course much higher.

Notation for Chapter 5

$\mathscr{A}, \mathscr{B}$	constants
c	intercept of equilibrium line on y-axis
G	flowrate of phase in which solute concentration is y
J_1	extraction factor for contactor 1 ($J_1 = m_1 S/L$)
J_2	extraction factor for contactor 2 ($J_2 = m_2 G/S$)
L	flowrate of phase in which solute concentration is x
m_1	slope of equilibrium line in contactor 1 (equation $z = m_1 x + c_1$)
m_2	slope of equilibrium line in contactor 2 (equation $y = m_2 z + c_2$)
N_R	number of pairs of cross-current extractor stages
N_{SO}	minimum number of stages in each contactor using counter-current forward and back extraction
Q_R	separation factor for cross-current extraction with recirculation ($Q_R = (y_{in} - y(x_{out}))/(y_{out} - y(x_{in}))$)
Q_O	optimum separation factor for counter-current forward and back extraction ($Q_O = (y_{in} - y(x_{out}))/(y_{out} - y(x_{in}))$)
S	flowrate of intermediate solvent (in which solute concentration is z)

x	solute concentration in phase of flowrate L
$x(y)$	x concentration in equilibrium with concentration y
y	solute concentration in phase of flowrate G ($x(y) = y/m_1m_2 - c_1/m_1 - c_2/m_1m_2$
$y(x)$	y concentration in equilibrium with concentration x ($y(x) = m_1m_2x + m_2c_1 + c_2$)
z	solute concentration in intermediate solvent (of flowrate S)

Greek symbol

β	overall extraction factor for crosscurrent extraction with recirculation ($\beta = (J_2 + J_1J_2)/(1 + J_2)$)

Subscripts

b	refers to intermediate solvent leaving contactor 2
G	refers to phase of flowrate G
in	refers to inlet of contactor
L	refers to phase of flowrate L
O	refers to optimum conditions in counter-current forward and back extraction
out	refers to outlet of contactor
R	refers to cross-current extraction with recirculation
t	refers to intermediate solvent leaving contactor 1
1	refers to contactor 1
2	refers to contactor 2

Superscript

*	denotes inversion

The above quantities may be expressed in any set of consistent units in which force and mass are not defined independently.

References

1. D. McNeil, Coal Tar Research Association, Private communication.
2. H. S. Mickley, T. K. Sherwood and C. B. Reed, *Applied Mathematics in Chemical Engineering*, McGraw-Hill Book Co. Inc., New York, 1957.
3. M. Souders and G. C. Brown, *Ind. Eng. Chem.* **24**, 519 (1932).
4. A. Kremser, *Nat. Pet. News* **22** (21), 42 (1930).

APPENDIX 5a
DESIGN AND OPERATION OF A CROSS-CURRENT FORWARD AND BACK EXTRACTOR

Figures A 5a.1–5 give the variation of N_R with Q_R for different values of J_2 at $J_1J_2 = 0{\cdot}1$, $0{\cdot}3$, $0{\cdot}5$, $0{\cdot}7$ and $0{\cdot}9$. They were obtained from equation 5(10) which yields straight line plots of N_R versus $\ln Q_R$ when J_1J_2 and J_2 are kept constant. In design, knowing the overall operating conditions J_1J_2 ($= m_1m_2G/L$) and the desired separation Q_R enables the required numbers of stages N_R to be determined for different solvent recirculation rates S (which fixes $J_2 = m_2G/S$). In operation knowing N_R and J_1J_2 and the desired separation Q_R enables J_2 (and hence S) to be determined.

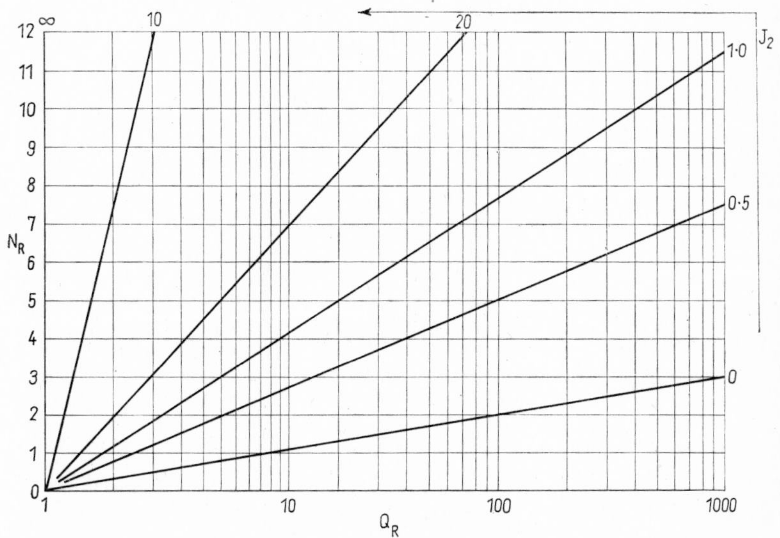

FIG. A 5a.1. N_R versus Q_R, for different values of J_2 at constant $J_1J_2 = 0{\cdot}1$.

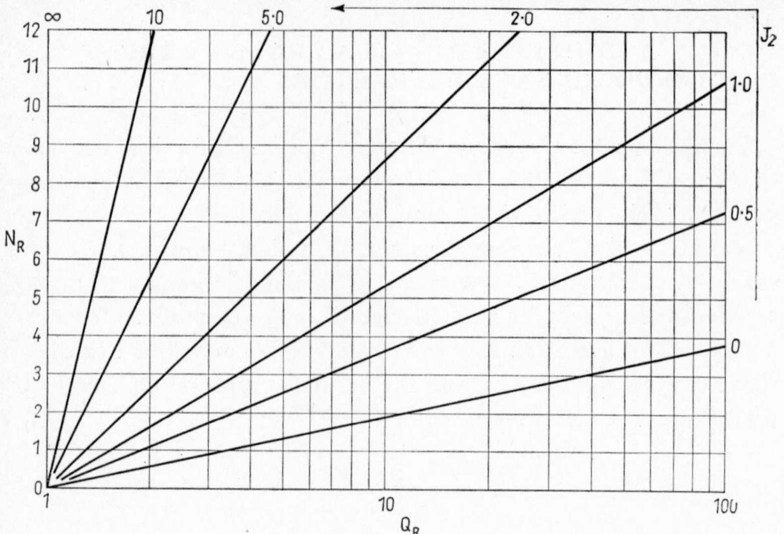

FIG. A5a.2. N_R versus Q_R, for different values of J_2 at constant $J_1J_2 = 0{\cdot}3$.

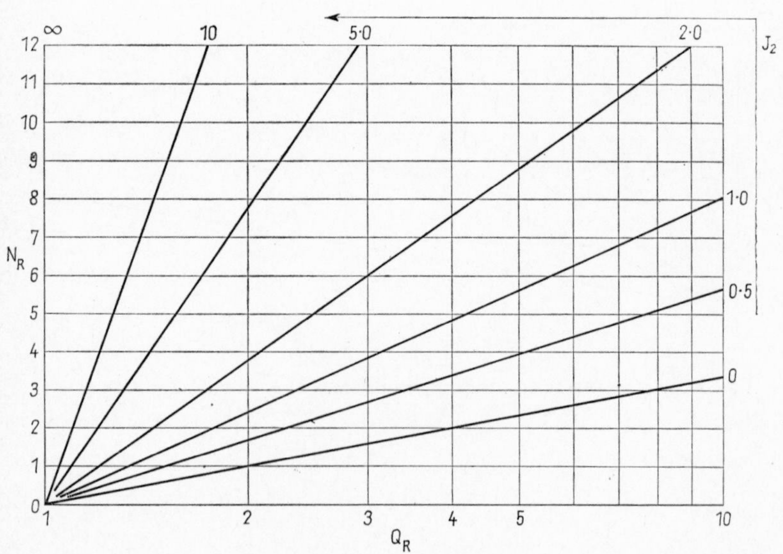

FIG. A5a.3. N_R versus Q_R, for different values of J_2 at constant $J_1J_2 = 0{\cdot}5$.

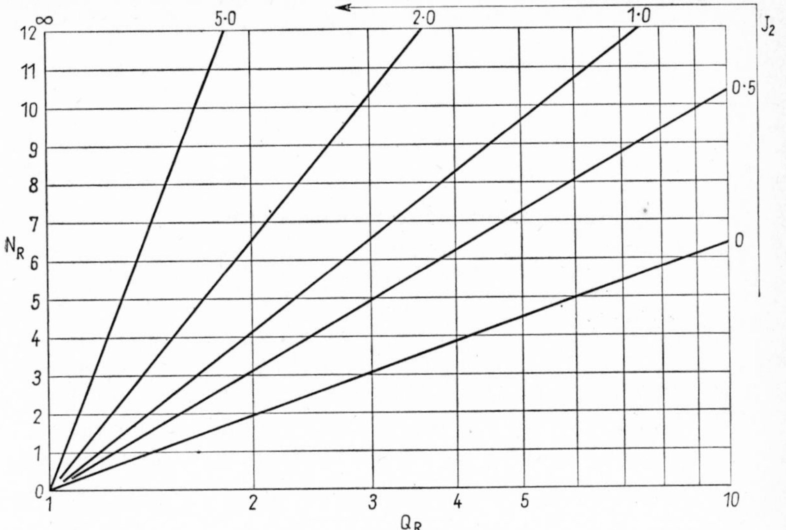

FIG. A 5a.4. N_R versus Q_R, for different values of J_2 at constant $J_1J_2 = 0{\cdot}7$.

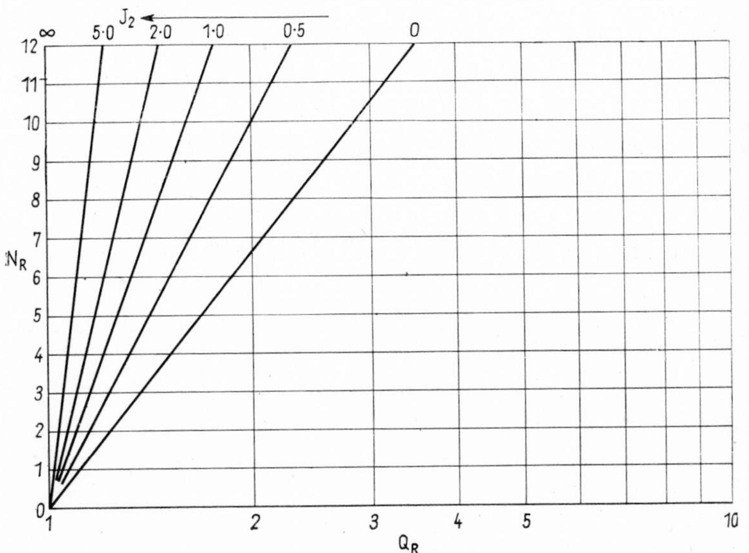

FIG. A 5a.5. N_R versus Q_R, for different values of J_2 at constant $J_1J_2 = 0{\cdot}9$.

CHAPTER 6

Linear Equilibrium Data

6.1. Introduction

The whole art of solvent extraction is based upon the study of liquid–liquid equilibrium. Many systems containing three or more components have been investigated and the experimental distribution data are reported in the periodical literature. These investigations have concentrated on the study of the effect of concentration on the distribution equilibrium. It has been observed that the ratio of the concentrations of a component in the two liquid phases in general varies with the concentration. Various laws of distribution have been proposed but these are frequently empirical[1,2] and no general rule is obeyed in practice over wide concentrations. Both mathematical expressions and graphical methods are employed to represent the distribution data for ternary systems. However, graphical representation is usually necessary for experimental distribution data. Simple mathematical expressions are obtained only as limiting cases, for example in the case of very dilute solutions, or when the systems contain two almost immiscible components. Figure 6.1 shows the shape of a typical distribution curve for ternary liquid systems. The components dominant in each phase are termed solvents and the third component distributed between the two phases is called the solute.

Many authors have in the past assumed linear equilibrium to ease mathematical computation in the analysis of extraction problems (e.g. refs. 3, 4, 5, 6) and it is important to know how applicable this assumption is. Furthermore, research workers studying various aspects of solvent extraction prefer to use systems in which the equilibrium is linear (e.g. refs. 7, 8). Some 150 systems were examined to discover how linear they are. The search was not exhaustive but limited to systems containing fairly common substances in which the solvents were reasonably immiscible. Referring again to Fig. 6.1, it is evident that in general we cannot fit a straight line through the origin to represent the whole concentration range without much error. However, two different straight lines may be fitted: one in the region *OA* and passing through

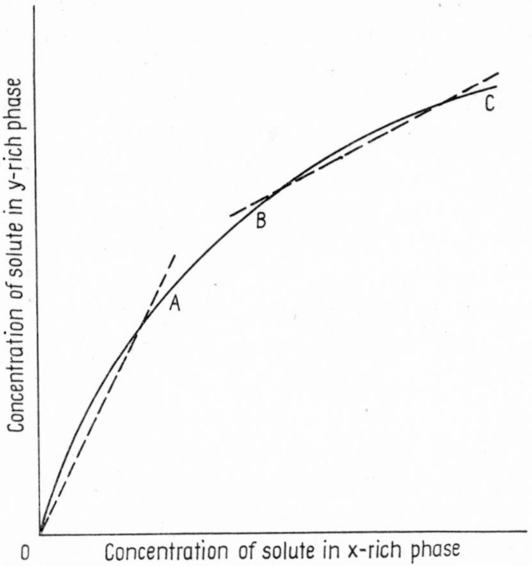

FIG. 6.1. A typical distribution curve.

the origin, and the other in the region *BC*. The slope and error of these two straight lines may be determined by a least squares procedure as described later.

Gas–liquid solubility data may be handled in a similar way. It is convenient to express the distribution in terms of Henry's Law

$$y = \mathscr{H}x$$

where $\mathscr{H}$ is a constant, y is the partial pressure of the solute in the gaseous phase and x its concentration in the liquid phase. These are usually reported as two-component systems (solute and liquid solvent) since a gaseous solvent is not essential to the distribution. The shape of a typical partial pressure versus concentration curve closely resembles

Fig. 6.1. A number of gas–liquid systems were thus investigated in the same way as the ternary liquid systems.

It is important to emphasise that the use of linearised data is not suitable for every purpose. Where careful design is required it is best to fit curves of higher orders and to do the arithmetic on a computer. This is particularly true when one is interested in non-linear portions of the equilibrium curve. An error of only 1% in the linearisation of the data may lead to as much as say 10% error in the design. Finally, isothermal conditions of operation are implied, and this may not always be so, especially in the case of gas absorption.

6.2. Theory

Minimising the sum of the squares of the errors in the x and/or y concentrations gives the best straight line through a series of x, y values as

$$y - \bar{y} = m(x - \bar{x}) \qquad 6(1)$$

where m is the slope of the line and $\bar{x}$, $\bar{y}$ are the arithmetic means of the x and y values respectively.

Assuming no error in the x concentrations and minimising the sum of the squares of the errors in the y concentrations as shown in Appendix 6a, the slope is given by

$$m_y = \frac{\Sigma x(y - \bar{y})}{\Sigma x(x - \bar{x})} \qquad 6(2)$$

the summations being over the p pairs of values. The variance in the slope is given by

$$e_{m_y}^2 = \frac{e_y^2}{\Sigma(x - \bar{x})^2} \qquad 6(3)$$

where e_y^2 is the variance in y, which is given by

$$e_y^2 = \frac{\Sigma((y - \bar{y}) - m_y(x - \bar{x}))^2}{p - 2} \qquad 6(4)$$

The variance in the mean value of y is given by

$$e_{\bar{y}}^2 = \frac{e_y^2}{p} \qquad 6(5)$$

Assuming no error in the y concentrations and minimising the sum of the squares of the errors in the x concentrations as shown in Appendix 6b, the slope is given by

$$m_x = \frac{\Sigma y\,(y - \bar{y})}{\Sigma y\,(x - \bar{x})} \qquad 6(6)$$

and the variance in the slope by

$$e_{m_x}^2 = \frac{m_x^4 e_x^2}{\Sigma\,(y - \bar{y})^2} \qquad 6(7)$$

where

$$e_x^2 = \frac{\Sigma\,((x - \bar{x}) - (1/m_x)\,(y - \bar{y}))^2}{p - 2} \qquad 6(8)$$

The variance in the mean value of x is given by

$$e_{\bar{x}}^2 = \frac{e_x^2}{p} \qquad 6(9)$$

Furthermore, it can be shown that

$$\frac{e_{m_x}}{m_x} = \frac{e_{m_y}}{m_y} \qquad 6(10)$$

and so it is only necessary to compute one of the fractional errors in the slope as discussed in Appendix 6f.

In general both the x and y concentrations will be in error and the average of the two slopes m_y and m_x should represent more accurately the "true" slope of the line. If, however, the variances e_x^2 and e_y^2 are known the functional relationship between x and y may be found. Lindley[(9)] in fact has derived an expression for any number of variates. Davies[(10)] summarizes the result for two variates where the ratio k^2 of the variances is known. The slope of the line is then given by

$$m = b + \sqrt{(b^2 + k^2)} \qquad 6(11)$$

where

$$b = \frac{\Sigma (y - \bar{y})^2 - k^2 \Sigma (x - \bar{x})^2}{2\Sigma (x - \bar{x})(y - \bar{y})} \qquad 6(12)$$

and

$$k^2 = e_y^2 / e_x^2 \qquad 6(13)$$

The variance in m is given by

$$e_m^2 = \frac{(k^2 + m^2)^2 \, e_y^2 / k^2}{k^2 \Sigma (x - \bar{x})^2 + m\Sigma (x - \bar{x})(y - \bar{y})} \qquad 6(14a)$$

or

$$e_m^2 = \frac{(k^2 + m^2)^2 \, e_x^2}{k^2 \Sigma (x - \bar{x})^2 + m\Sigma (x - \bar{x})(y - \bar{y})} \qquad 6(14b)$$

where the variances in x or y may be estimated from

$$e_y^2 = \frac{\Sigma (y - \bar{y})^2 - m\Sigma (x - \bar{x})(y - \bar{y})}{p - 2} \qquad 6(15a)$$

or

$$e_x^2 = \frac{\Sigma (x - \bar{x})^2 - (1/m) \Sigma (x - \bar{x})(y - \bar{y})}{p - 2} \qquad 6(15b)$$

The variances in the mean values may be estimated from

$$e_{\bar{y}}^2 = k^2 e_{\bar{x}}^2 = e_y^2 / p = k^2 e_x^2 / p \qquad 6(16)$$

When there is no error in the x concentrations, k tends to infinity and equations 6(11) and (12) reduce to equation 6(2), equation 6(14a) to equation 6(3) and equation 6(15a) to equation 6(4) as shown in Appendix 6e. When there is no error in the y concentrations, k tends to zero and equations 6(11) and (12) reduce to equation 6(6), equation 6(14b) to equation 6(7) and equation 6(15b) to equation 6(8). The regression of y upon x is thus the limit of the functional relationship when k tends to infinity and the regression of x upon y the limit when k tends to zero.

Of course, if the variances e_x^2 and e_y^2 are known the value of k may be determined. Alternatively e_x^2 and e_y^2 may be estimated using equations 6(15a) and (15b). This involves a knowledge of the slope m, which in turn depends on k and so an iterative procedure results; however, the extra effort yields more accurate values of k and m.

All the above equations may be applied to fit straight lines in the upper concentration region of the distribution curve. In general these lines do not pass through the origin but to fit a straight line in the lower concentration region of the distribution curve, the same set of equations may be used, noting that the general form (given by equation 6(1)) reduces to

$$y = Mx \tag{6(17)}$$

when $\bar{x} = \bar{y} = 0$ and m is replaced by M to distinguish lines through the origin. We can make $\bar{x}$ and $\bar{y} = 0$ by postulating a set of points identical to the experimental set but reflected through the origin. Thus all the previous equations apply when we replace $\bar{x}$ and $\bar{y}$ by 0 and $(p - 2)$ by $(p - 1)$, (as only one degree of freedom is used up in estimating M). The equations for the two regressions through the origin are derived from first principles in Appendices 6c and 6d.

6.3. Experimental Data

6.3.1. Sources of Data

Various authors have published handbooks listing ternary liquid systems, giving references to the literature for the distribution data. In addition, some of these handbooks report these data. The following were found to be useful starting-points for a more comprehensive literature search:

(a) *The International Critical Tables.* The section of greatest interest is the one by Forbes and Anderson[11] on "Phase equilibrium data for condensed systems containing two liquid phases with a third component in distribution equilibrium between them, the two liquid phases being practically non-miscible". The systems are listed according to the solute and solvents. Aqueous systems are split into those containing organic and those containing inorganic consolute components. Distribution coefficients are given.

(b) Seidell has compiled considerable solubility data from the periodical literature in three handbooks. The first two published in 1940–1 compile the data available up to that date, dividing the organic and inorganic systems into two separate volumes. The volume of more interest is the one dealing with organic systems.[12] The compounds are listed alphabetically or by empirical formula. A certain amount of ternary equilibrium data is listed in both graphs and tables.

In the third volume published in 1952[13] Francis lists ternary systems separating into two liquid layers. Classification is made into aqueous and non-aqueous systems. Some distribution coefficients are given.

(c) Francis[14] has systematically tabulated a large number of liquid–liquid references up to *Chemical Abstracts* for May 1963. Classification is, as before, into aqueous and non-aqueous systems. Most of the references are to triangular equilibrium diagrams and distribution data, if given, are usually tabulated.

Data for gas–liquid systems were found mainly in the *International Critical Tables*,[15] and from this source were used directly. Additional data was obtained from Seidell[12] and the *Journal of Chemical and Engineering Data*.

6.3.2. *Analysis of Data*

A total of 184 systems were investigated being made up of ninety-nine aqueous–organic systems (AO), forty-seven organic–organic systems (OO) and thirty-nine liquid–gas systems (LG). These are arranged alphabetically according to the solute in Appendix 6g and some are listed in Table 6.1. Each separate set of data is termed a system and given a separate number and so there are included in the list identical physical systems with data obtained by different authors or at different temperatures. The concentration units in which the data are given are also tabulated together with the reference to their source.

As is usual in the literature the solute concentrations x and y refer to the two conjugate phases (which separate out after a mixture of the solute and two solvents have been agitated until equilibrium is reached). For aqueous–organic systems the water phase is for the sake of uniformity always designated as the x-phase and the other phase as the y-phase. For organic–organic systems the x- and y-phases are designated

TABLE 6.1. SYSTEMS WITH ERROR IN FULL RANGE SLOPE OF LESS THAN 1%

Aqueous–Organic Systems

System number	Solute	Organic solvent y-phase	Temp. °C	Concentration units	Reference
AO-10	Acetic acid	Methyl cyclohexanone	23	weight %	16
AO-32	n-Butyric acid	Ethyl butyrate	28	weight %	17
AO-38	n-Caproic acid	Methyl isobutyl ketone	20	g moles/l	18
AO-39	Diethylamine	Methyl isobutyl ketone	20	g moles/l	19
AO-40	Diethylamine	Toluene	20	g moles/l	19
AO-47	Formic acid	Methyl isobutyl ketone	25	g moles/l	18
AO-54	Lactic acid	l-Butanol	25	weight %	20
AO-56	Malonic acid	Ethyl ether	25	mg moles/l	21
AO-67	Nicotine	Benzene	25	g/l	22
AO-68	Nicotine	Carbon tetrachloride	25	g/l	22
AO-73	Oxalic acid	n-Amyl alcohol	Ambient	g/l	23
AO-83	Propionic acid	Methyl butyrate	30	weight %	24
AO-85	Propionic acid	Olive oil	25	g moles/l	25
AO-94	Succinic acid	n-Amyl alcohol	Ambient	g/l	23
AO-96	Triethylamine	Methyl isobutyl ketone	20	g moles/l	19
AO-99	n-Valeric acid	Methyl isobutyl ketone	25	g moles/l	18

Organic–Organic Systems

System number	Solute	Solvent x-phase	Solvent y-phase	Temp. °C	Concentration units	Reference
OO-24	Diphenylhexane	Docosane	Furfural	45	weight %	26

Liquid–Gas Systems

System number	Solute gas y-phase	Liquid x-phase	Temp. °C	Concentration units		Reference
				y-phase	x-phase	
LG-01	Ammonia	Water	0	mm Hg (partial)	g/g	27
LG-02	Ammonia	Water	20	mm Hg (partial)	cm^3/cm^3*	
LG-13	Carbon Dioxide	Water	20	atmospheres (total)	cm^3/cm^3*	28
LG-14	Carbon dioxide	Water	35	atmospheres (total)	cm^3/cm^3*	
LG-15	Carbon dioxide	Water	35	atmospheres (total)	cm^3/cm^3*	
LG-17	Carbon dioxide	Water	60	atmospheres (total)	cm^3/cm^3*	
LG-18	Carbon dioxide	Water	100	atmospheres (total)	cm^3/cm^3*	
LG-26	Propylene	Water	37.8	psia (total)	mol %	29
LG-33	Dimethyl ether	Acetone	25	mm Hg (partial)	mole fraction	30
LG-34	Dimethyl ether	Benzene	25	mm Hg (partial)	mole fraction	30
LG-35	Dimethyl ether	Carbon tetra-chloride	25	mm Hg (partial)	mole fraction	30

* cm^3/cm^3 refers to cubic centimetres of gas (reduced to STP) dissolved in one cubic centimetre of liquid solvent.

arbitrarily but are made explicit in Appendix 6g and Table 6.1. The liquid phase is always taken as the x-phase in liquid–gas systems.

Using the formulae of the previous section the x and y concentrations so obtained were used to calculate the two lines shown in Fig. 6.1 having the minimum error in their slope. The slope of the best straight line passing through the origin was obtained by taking the points in order of increasing concentration. Starting with the two points nearest the origin the slope was obtained and then further points were successively added until all the points were included, the slope and its error being determined in each case. The slope of the other straight line, through the mean values $\bar{x}$ and $\bar{y}$, was obtained by taking the points in decreasing order of concentration and starting with the three points farthest away from the origin. A digital computer program was used to perform the calculation as described in Appendix 6f.

6.4. Results and Discussion

Least Squares Criteria

Three kinds of straight lines have been considered: a regression assuming no error in x, another assuming no error in y, and one assuming a functional relationship in which the ratio of the errors in x and y is k. To determine how the functional relationship behaves with varying k, equations 6(11) and (14) giving the slope and the error in the slope were evaluated for a range of values of k. The results for the slope m of system AO-01 are plotted in Fig. 6.2. As k tends to zero m tends to m_x, and as k tends to infinity m tends to m_y. The function is well behaved for intermediate values of k. Thus we may conclude that the extreme values of the slope given by the functional relationship are those of the two regressions. Furthermore, percentage errors in the slope were found to be equal as k approached zero and infinity, the value being that obtained by the other two regressions. The percentage error passes through a minimum at $k = 1$.

For each system the slopes were compared for the full composition range obtained by the different least squares criteria. The slope m was obtained for $k = 1$, and the value of m was always between m_y and m_x, being close to their average. Also the errors in the slopes were obtained and the percentage error in m (for $k = 1$) is always less than that for the two regressions. Similar results were obtained for the slope M. All these observations tally with what we have concluded in the preceding

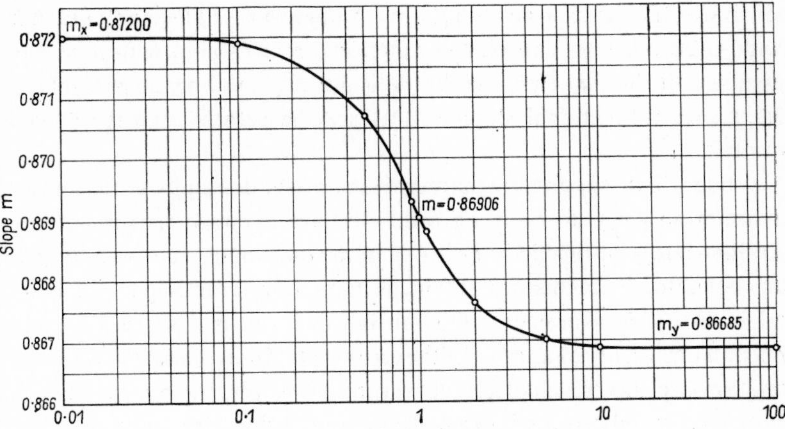

FIG. 6.2. Variation of slope m with ratio of standard deviations k in functional relationship for system AO-01. Acetaldehyde between water and vinyl acetate at 20°C.

paragraph. We may thus safely say that the regression of y upon x (no error in x) is the limit of the functional relationship when k tends to infinity and the regression of x upon y (no error in y) is the limit when k tends to zero, as deduced in the theoretical section.

Before we can tabulate the best straight lines through the experimental data we must decide which least squares criterion to apply. We have seen how the functional slope takes on values intermediate to those given by the two regressions. In most cases the difference between these two extreme slopes is small and so it would appear safe to use the functional slope with $k = 1$. With no evidence to the contrary it seems reasonable to suppose that measurements in the x and y concentrations for liquid–liquid systems are subject to equal error. This is not the case for the gas–liquid systems where two entirely different quantities are measured. However, we have already seen that whatever the value of k, the functional slope is well behaved and varies only within the narrow limits set by the two regressions. Since we do not know the true value for k for the gas–liquid systems it is best for simplicity to assume, as for the liquid–liquid systems that $k = 1$. Estimates of the two variances may be sometimes obtained from the original references but this is not always so. Therefore when tabulating the best straight lines in Table 6.3 and Appendices 6h and 6i, the functional relationship with $k = 1$ is assumed. It should be noted that if k is not equal to one the percentage error in the functional slope will always be greater than the values quoted.

While still discussing the mathematical methods employed one basic assumption needs to be examined. We have assumed that there is a straight-line relationship in portions of the distribution curve, and that any deviation from this straight line is due solely to random experimental errors of measurement. Obviously this is not true when the equilibrium line is curved as most of the error in assuming a linear functional relationship then results from curvature. However, these methods are employed merely to distinguish the degree of linearity and they need not be rigorous in every aspect. Too much significance should not be attached to the slopes and errors quoted for the more non-linear systems. Many of the systems investigated are fairly linear and as the linearity increases so the assumption becomes more valid.

Tabulation

The parameters of interest are the slopes, mean values and errors in these quantities. From the results of the computer output it is evident that the error in the slope usually passes through a minimum as the number of points increases (this is especially so when the line is curved). The number of systems having an error in the full range slope, and a minimum error, of between 0–1 %, 1–2 %, 2–4 % and greater than 4 % are given in Table 6.2. To enable the reader to visualise what a 1 %, 2 % or 4 % error in the slope involves, a few systems are plotted in Figs. 6.3 to 6.6 and the fitted lines drawn on them. From such results it is concluded that less than 1 % error represents a very linear system, less than 2 % error a fairly linear system and greater than 4 % a non-linear system. Table 6.2 shows that 16 % of aqueous–organic systems,

2% of organic–organic systems and 28% of liquid–gas systems have a full range error of less than 1% when the functional relationship is expressed as $y = Mx$ or $y - \bar{y} = m(x - \bar{x})$. However, 38% of aqueous–organic systems, 26% of organic–organic systems and 77%

TABLE 6.2. NUMBER OF SYSTEMS HAVING ERROR IN SLOPE WITHIN LIMITS SHOWN

System	Full range error				Minimum error			
	0–1%	1–2%	2–4%	>4%	0–1%	1–2%	2–4%	>4%
	$y = Mx$							
AO	13	13	27	46	33	19	22	25
OO	1	5	9	32	11	11	13	12
LG	6	5	13	15	15	4	6	14
	$y - \bar{y} = m(x - \bar{x})$							
AO	10	10	18	61	13	19	20	47
OO	0	1	6	40	2	1	6	38
LG	7	9	9	14	23	3	5	8
	$y = Mx$ or $y - \bar{y} = m(x - \bar{x})$							
AO	16	12	30	41	38	21	24	16
OO	1	6	8	32	12	12	10	13
LG	11	7	11	10	30	1	2	6

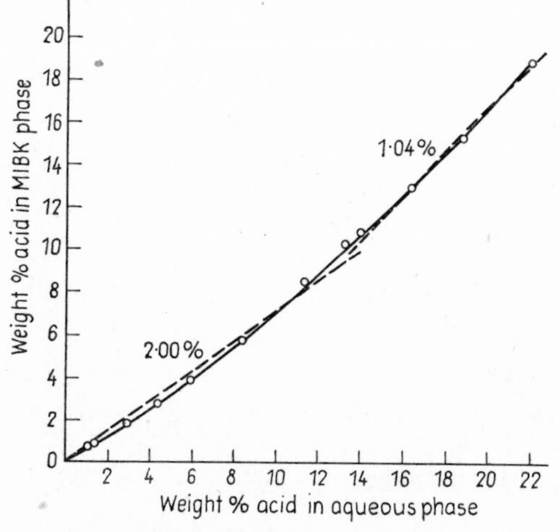

FIG. 6.3. Distribution curve for system AO-14. Acetic acid between water and methyl isobutyl ketone at 28 °C.

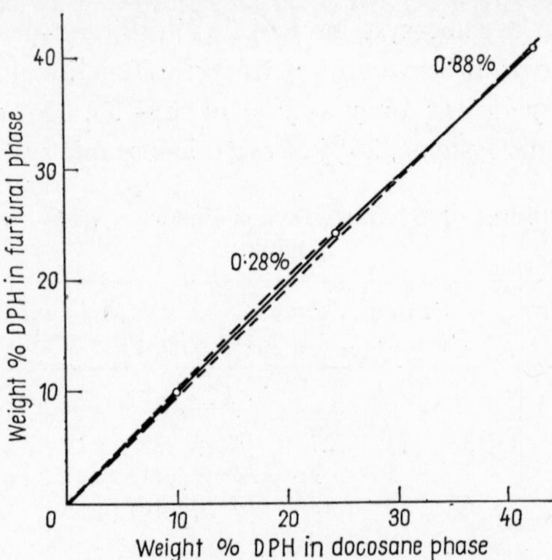

FIG. 6.4. Distribution curve for system OO-24. Diphenyl hexane between docosane and furfural at 45 °C.

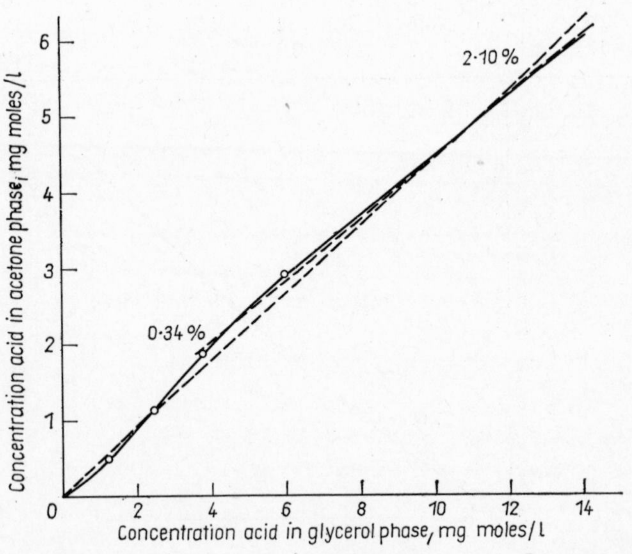

FIG. 6.5. Distribution curve for system OO-32. Malonic acid between glycerol and acetone at 25 °C.

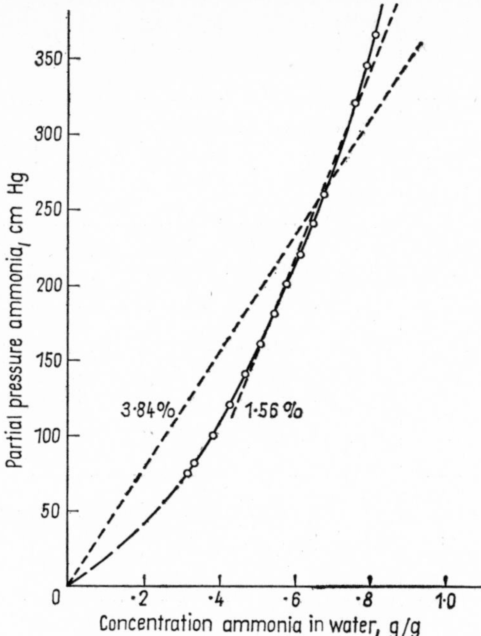

FIG. 6.6. Distribution curve for system LG-11. Ammonia in water at 40°C.

of liquid–gas systems possess a smaller region in which the error is less than 1%.

Appendices 6h and 6i summarise the parameters which are sufficient fully to characterise the systems. Appendix 6h lists the parameters for the line in the lower concentration range ($y = Mx$) and Appendix 6i for the line in the upper concentration range $(y - \bar{y}) = m(x - \bar{x})$. For reasons of space and clarity it is evident the parameters cannot be listed for each system, still less for every point. Results are only given for those systems having a region for which the error is less than 4%. Three concentration ranges are quoted; the full range, the range over which the error in the slope is least, and the widest range over which the error in the slope does not exceed about 2%. The latter working range is, of course, identical with that having the minimum error when this exceeds 2%.

The parameters for those systems having an error in the slopes of less than 1% over the full range are given in Table 6.3 and when there

TABLE 6.3. PARAMETERS OF SYSTEMS WITH ERROR IN FULL RANGE SLOPE OF LESS THAN 1%

Lower concentration region			System number	Upper concentration region					
Upper concentration x-phase	Slope M	% error in M		Lower concentration x-phase	$\bar{x}$	$\bar{y}$	Error in $\bar{x}$ or $\bar{y}$	Slope m	% error in m
31	0·979	0·33	AO-10	5·45	18·91	18·48	0·049	0·985	0·76
				15·15	22·28	21·83	0·029	0·973	0·58
6·5	6·635	0·73	AO-32	3·0	4·967	32·833	0·020	7·024	1·65
				4·7	5·700	38·075	0·006	6·588	0·92
0·01125	79·406	0·95	AO-38	0·00553	0·0084	0·6640	0·00002	85·467	1·05
0·1842	0·663	0·85	AO-39	0·00454	0·05350	0·03426	0·00016	0·688	0·57
				0·01068	0·06036	0·03880	0·00015	0·692	0·54
0·590	0·557	0·61	AO-40	0·00099	0·0717	0·0383	0·0002	0·564	0·36
				0·00629	0·754	0·0403	0·0002	0·565	0·35
2·678	0·445	0·72	AO-47	0·170	0·9999	0·4393	0·0026	0·453	0·81
0·464	0·421	0·37							
12·5	0·928	0·75	AO-54	0·52	7·484	6·903	0·036	0·951	1·37
10·3	0·919	0·56		9·60	10·800	10·123	0·0004	1·045	0·05
94·6	0·148	0·59	AO-56	2·19	27·535	4·078	0·040	0·1487	0·86
3·8	0·135	0·19		22·30	50·440	7·537	0·018	0·1463	0·38
1·87	10·317	0·29	AO-67	0·45	1·126	11·586	0·002	10·446	0·56
1·65	10·276	0·23		0·61	1·238	12·748	0·002	10·499	0·48

1·97	9·551	0·73	AO-68	0·44	1·027	9·820	0·009	9·508	1·72
1·69	9·641	0·67							
81·6	0·362	1·09	AO-73	2·14	32·213	11·276	0·086	0·381	0·96
14·6	2·119	1·00	AO-83	6·2	10·750	22·850	0·095	2·025	3·74
11·8	2·159	0·63		11·8	13·200	27·867	0·013	1·536	1·34
0·218	0·166	0·38	AO-85	0·176	0·1077	0·0179	0·0001	0·1662	0·75
67·5	0·607	0·96	AO-94	2·81	27·369	16·883	0·136	0·590	1·28
25·5	0·642	0·27							
0·02947	5·894	1·01	AO-96	0·00135	0·01447	0·08288	0·00005	6·305	0·53
				0·00519	0·01646	0·09515	0·00003	6·376	0·41
0·0358	25·143	0·94	AO-99	0·0198	0·0279	0·6988	0·0001	26·945	1·84
0·0255	24·568	0·79							
42·6	0·968	0·88	OO-24	10·0	25·700	24·967	0·193	0·953	2·10
24·5	0·987	0·28							
1·887	951·2	0·77	LG-01	0·997	1·476	1397	0·005	10·58	1·60
1·688	935·0	0·53		1·778	1·837	1788	0·001	1508	0·00
2483	0·722	0·77	LG-02	1312	1942	1397	3·87	0·803	1·60
2221	0·710	0·54							
26·87	1·844	2·98	LG-13	16·33	21·894	39·714	0·010	2·641	0·30
23·43	2·946	0·40	LG-14	9·98	16·851	49·778	0·031	2·834	0·75
13·38	2·996	0·15		20·39	22·033	64·333	0·004	2·633	0·37
29·6	2·614	0·23	LG-15	11·2	20·891	54·727	0·010	2·551	0·19
13·1	2·675	0·13							

TABLE 6.3 (*cont.*)

Lower concentration region				Upper concentration region					
Upper concentration x-phase	Slope M	% error in M	System number	Lower concentration x-phase	$\bar{x}$	$\bar{y}$	Error in $\bar{x}$ or $\bar{y}$	Slope m	% error in m
26·6	4·60	0·98	LG-17	8·5	17·188	79·765	0·121	4·203	2·18
16·6	4·817	0·21		24·7	25·867	113·667	0·007	3·147	0·88
16·15	9·935	0·35	LG-18	6·35	11·127	110·130	0·027	10·377	0·84
15·91	9·908	0·32		6·84	11·576	115·000	0·023	10·241	0·77
0·09506	2415	0·98	LG-26	0·01122	0·05357	127·09	0·00040	2585	1·56
0·1893	6413	6·86	LG-33	0·0179	0·0978	670·03	0·0005	4456	0·81
				0·0378	0·1111	729·75	0·0000	4442	0·00
0·210	4994	2·72	LG-34	0·023	0·1219	626·91	0·0000	4327	0·00
0·233	4828	3·20	LG-35	0·0300	0·1274	637·61	0·0000	4106	0.00
				0·1633	0·1985	929·33	0·0000	4168	0·00

is a smaller concentration range over which the error is less this is also given. The error in M is usually less than that in m even though the latter represents the slope of the best straight line through the experimental points. This is because the line of slope M passes through the origin whereas the line of slope m passes through the mean value of the observations. It is thus possible for the error in x or y estimated from $y = Mx$ to be greater than that estimated from $y - \bar{y} = m(x - \bar{x})$ over the greater part of the concentration range, even though the error in M is less than that in m.

Extraction Studies

Smoot and Babb[7] and Vermijs and Kramers[8] have used the system acetic acid–water–methyl isobutyl ketone for extraction studies and have assumed linear equilibrium lines. Smoot has averaged the available equilibrium data in the temperature range 20–30 °C and obtained a slope of 0·52 up to a concentration of 12×10^{-3} lb moles acetic acid/ft^3 of water. This slope when translated into weight per cent units becomes 0·61 and the concentration limit 1·2 wt. %. The slope obtained for a similar range in system AO-13 using weight per cent units is about 0·54 and varies from about 0·49 to 0·62 as the concentration in the x-phase increases from about 0·5 to 2·9 wt. % (see Appendix 6h, system AO-13). Vermijs and Kramers acknowledge a slight variation in distribution coefficient up to an aqueous concentration of about 7 wt. % and used the data of system AO-14. This is the same physical system as AO-13 but at a slightly elevated temperature, and extends over a higher concentration range. Figure 6.3 is a plot of this system and shows that the value of the distribution coefficient is 0·62 up to a concentration of 1·4wt. % (which agrees with Smoot's figure). At much higher concentrations the value of the distribution coeficent is 0·80. In spite of this variation the errors encountered are of the order 1–3 %, provided the correct distribution coefficients are used over any given range.

Physical Properties

As a further guide to the choice of an experimental system the physical properties and other characteristics of most of the solvents and solutes of the systems investigated are listed in Appendix 6j. Most of the data were obtained from Sax[31] and Perry[32]. The flash points measured by the open and/or closed cup methods are given, together with the explosive concentration range of the vapour in air, the normal boiling point, freezing point, viscosity at 20 °C and the specific gravity. The toxicity ratings correspond roughly to those given in Sax though these can only be taken as rough guides, since the nature and length of exposure to the material cannot be taken into account. The original text should be consulted for further details.

Notation for Chapter 6

b	defined by equation 6(12)
c	intercept on y-axis of line passing through $\bar{x}$, $\bar{y}$
D	sum of squares of deviations
e	standard deviation
e^2	variance
$\mathscr{H}$	Henry's law constant
k^2	ratio of variance in y to variance in x in functional relationship
m	slope of line passing through $\bar{x}$, $\bar{y}$
M	slope of line passing through origin
p	number of pairs of observations
x, y	concentrations of solute in two conjugate phases
$\bar{x}$	arithmetic mean of the x values
$\bar{y}$	arithmetic mean of the y values
X	best estimate of x for a given value of y
Y	best estimate of y for a given value of x

Greek symbol

Σ	indicates summation over total number of observations

Subscripts

i	refers to ith pair of observations
p	refers to pth pair of observations
x	refers to regression of x on y
y	refers to regression of y on x
1, 2 ...	refers to 1st, 2nd ... pairs of observations

Superscript

* denotes inversion

The above quantities may be expressed in any set of consistent units in which force and mass are not defined independently.

References

1. J. C. ELGIN and F. M. BROWNING, Extraction of acetic acid with isopropyl ether in a spray column, *Trans. Am. Inst. Chem. Engrs.* **31,** 639 (1935).
2. L. L. SIMS and D. W. BOLME, Liquid–liquid equilibria for the system ethylene glycol–toluene–acetone, *J. Chem. Eng. Data* **10,** 111 (1965).
3. A. P. COLBURN, The simplified calculation of diffusional processes. General consideration of two film resistances, *Trans. Am. Inst. Chem. Engrs.* **35,** 211 (1939).
4. A. KREMSER, Theoretical analysis of absorption process, *Nat. Petr. News* **22,** 42 (1930).
5. A. KLINKENBERG, Calculation of the efficiency of counter-current stagewise mass transfer processes with constant distribution factor when in the steady state. I. Distribution of one component only, *Chem. Eng. Sci.* **1,** 86 (1951).
6. S. HARTLAND and J. C. MECKLENBURGH, A comparison of differential and stagewise counter-current extraction with backmixing. *Chem. Eng. Sci.* **21,** 1209 (1966).
7. L. D. SMOOT and A. L. BABB, Mass transfer studies in a pulsed extraction column. Longitudinal concentration profiles, *Ind. Eng. Chem. (Fundamentals)* **1,** 93 (1962).
8. H. J. A. VERMIJS and H. KRAMERS, Liquid extraction in a rotating disc contactor, *Chem. Eng. Sci.* **3,** 56 (1954).
9. D. V. LINDLEY, Regression lines and the linear functional relationship, *Supplement J. Royal Statistical Soc.* **9,** 218 (1947).
10. O. L. DAVIES, *Statistical Methods in Research and Production*, 3rd ed., pp. 173–4, Oliver & Boyd, Edinburgh, 1957.
11. G. S. FORBES, G. W. ANDERSON, A. E. HILL and A. S. COOLIDGE, *International Critical Tables*, Vol. 3, pp. 386–433, McGraw-Hill, New York, 1928.
12. A. SEIDELL, *Solubilities of Inorganic Compounds*, Vol. 2, 3rd ed., Van Nostrand, Princeton, New Jersey, 1941.
13. A. W. FRANCIS, *Solubilities of Inorganic and Organic Compounds*, supplement to the 3rd ed., A. SEIDELL and W. R. LINKE (Editors), *Ternary Systems Separating into Two Liquid Layers*, Van Nostrand, 1952.
14. A. W. FRANCIS, *Liquid–Liquid Equilibriums*, Interscience, New York, 1963.
15. A. G. LOOMIS, Solubilities of gases in liquids, *International Critical Tables*, Vol. 3, pp. 255–70, McGraw-Hill, New York, 1928.
16. D. F. OTHMER, R. E. WHITE and E. TRUEGER, Liquid–liquid extraction data, *Ind. Eng. Chem.* **33,** 1240 (1941).
17. M. V. R. RAO and P. D. MURTY, Ternary liquid equilibria, *J. Chem. Eng. Data* **10,** 248 (1965).

18. E. Bak and C. J. Geankoplis, Distribution coefficients of fatty acids between water and methyl isobutyl ketone, *Ind. Eng. Chem.—Chem. Eng. Data Series* **3,** 256 (1958).
19. K. C. Williams and S. R. M. Ellis, Equilibrium distribution ratios of diethylamine and triethylamine at low concentrations, *J. Appl. Chem.* **11,** 492 (1961).
20. V. E. Petrits and C. J. Geankoplis, Phase equilibria in 1-butanol–water–lactic acid system, *J. Chem. Eng. Data* **4,** 197 (1959).
21. H. W. Smith, The nature of secondary valence, *J. Phys. Chem.* **25,** 616 (1921).
22. R. T. Fowler and R. A. S. Noble, Distribution experiments and liquid–liquid equilibrium in systems containing nicotine, *J. Appl. Chem.* **4,** 546 (1954).
23. C. F. Asselin and E. W. Comings, Fractional solvent extraction with reflux, *Ind. Eng. Chem.* **42,** 1198 (1950).
24. N. S. Murty, V. Subrahmanyam and P. D. Murty, Ternary liquid equilibria, *J. Chem. Eng. Data* **11,** 335 (1966).
25. M. Bodansky and A. V. Meigs, The distribution ratios of some fatty acids and their halogen derivatives between water and olive oil, *J. Phys. Chem.* **36,** 814 (1932).
26. S. W. Briggs and E. W. Comings, Effect of temperature on liquid–liquid equilibrium: benzene–acetone water system and docosane-1,6-diphenylhexane–furfural system, *Ind. Eng. Chem.* **35,** 411 (1943).
27. B. S. Neuhausen and W. A. Patrick, A study of the system ammonia and water as a basis for the theory of the solution of gases in liquids, *J. Phys. Chem.* **25,** 693 (1922).
28. W. Sander, The solubility of carbon dioxide in water and some other solvents under high pressure, *Zeitschrift für Physik Chemie* **78,** 513 (1911).
29. A. Azarnoosh and J. J. McKetta, Solubility of propylene in water, *J. Chem. Eng. Data* **4,** 211 (1959).
30. J. Horiuti, On the solubility of gas and coefficient of dilatation by absorption, *Sci. Papers Inst. Phys. Chem. Res. (Tokyo)* **17,** 125 (1931).
31. N. I. Sax, *Dangerous Properties of Industrial Materials*, Reinhold, New York, 1957.
32. J. H. Perry *et al.* (Editors), *Chemical Engineers' Handbook*, 4th ed., McGraw-Hill, New York, 1963.
33. R. C. Pratt and S. T. Glover, Liquid–liquid extraction: removal of acetone and acetaldehyde from vinyl acetate with water in a packed column, *Trans. Inst. Chem. Engrs. (London)* **24,** 54 (1946).
34. A. E. Skrzec and N. F. Murphy, Liquid–liquid equilibria, acetic acid in water with 1-butanol, methyl ethyl ketone, furfural, cyclohexanol and nitromethane, *Ind. Eng. Chem.* **46,** 2245 (1954).
35. P. Eaglesfield, B. K. Kelly and J. F. Short, Recovery of acetic acid from dilute aqueous solutions by liquid–liquid extraction—Part 2, *Ind. Chemist* **29,** 243 (1953).
36. E. L. Heric and R. M. Routledge, Distribution of acetic and propionic acids between furfural and water, *J. Chem. Eng. Data* **5,** 272 (1960).
37. J. C. Upchurch and M. Van Winkle, Liquid–liquid equilibria. Heptadecanol–water–acetic acid and heptadecanol–water–ethanol, *Ind. Eng. Chem.* **44,** 618 (1952).
38. K. Murali and M. R. Rao, Mass transfer studies in perforated plate towers, *J. Chem. Eng. Data* **7,** 468 (1962).

39. E.G. SCHEIBEL and A.E. KARR, Semi-commercial multistage extraction column—performance characteristics, *Ind. Eng. Chem.* **42,** 1048 (1950).
40. K. ISHII, S. HAYANI, T. SHIRAI and K. ISHIDA, Liquid equilibrium data for the system propane, propylene and ammonia solvents, *J. Chem. Eng. Data* **11,** 288 (1966).
41. H.F. JOHNSON and H. BLISS, Liquid–liquid extraction in spray towers, *Trans. Am. Inst. Chem. Engrs.* **42,** 331 (1946).
42. S.B. ROW, J.H. KOFFOLT and J.R. WITHROW, Characteristics and performance of a nine-inch liquid–liquid extraction column, *Trans. Am. Inst. Chem. Engrs.* **37,** 559 (1941).
43. J.H. JONES and J.F. MCCANTS, Ternary solubility data: 1-butanol–methyl 1-butyl ketone–water, 1-butyraldehyde–ethyl acetate–water, 1-hexane–methyl ethyl ketone–water, *Ind. Eng. Chem.* **46,** 1956 (1954).
44. E.L. HERIC, B.H. BLACKWELL, L.J. GAISSERT, S.R. GRANT and J.W. PIERCE, Distribution of butyric acid between furfural and water at 25 and 35 °C, *J. Chem. Eng. Data* **11,** 38 (1966).
45. C.A. CHANDY and M.R. RAO, Ternary liquid equilibria l-hexanol–water–fatty acids, *J. Chem. Eng. Data* **7,** 473 (1962).
46. M.R. RAO, M. RAMAMURTY and C.V. RAO, Ternary liquid equilibria, *Chem. Eng. Sci.* **8,** 265 (1958).
47. D.G. BEECH and S. GLASSTONE, Solubility influence. Part V. The influence of aliphatic alcohols on the solubility of ethyl acetate in water, *J. Chem. Soc. (London)* **67** (1938).
48. S. HORIBA, Equilibrium in the system; water, ethyl alcohol and ethyl ether, *Mem. Coll. Sci. Eng. (Kyoto)* **3,** 63 (1911).
49. K.E. WHITEHEAD and C.J. GEANKOPLIS, Separation of formic and sulphuric acids by extraction, *Ind. Eng. Chem.* **47,** 2114 (1955).
50. H.W. SMITH and T.A. WHITE, The distribution ratios of some organic acids between water and organic liquids, *J. Phys. Chem.* **33,** 1953 (1929).
51. N.A. KOLOSOVSKII, A. BEKTUROV and F.S. KULIKOV, Distribution of saturated monobasic aliphatic acids between two immiscible liquid phases, *J. Gen. Chem. (U.S.S.R.)* **5,** 319 (1935).
52. N.A. KOLOSOVSKII, N.A. KULIKOV and A. BEKTUROV, Distribution of isovaleric acid between two liquid phases (XI), *J. Gen. Chem. (U.S.S.R.)* **4,** 1153 (1934).
53. R.B. WEISER and C.J. GEANKOPLIS, Lactic acid purification by extraction, *Ind. Eng. Chem.* **47,** 858 (1955).
54. R.J. RAO and C.V. RAO, Ternary liquid equilibria: methanol–water–esters, *J. Appl. Chem.* **7,** 435 (1957).
55. E.R. WASHBURN and H.C. SPENCER, A study of solutions of methyl alcohol in cyclohexane, in water and in cyclohexane and water, *J. Am. Chem. Soc.* **56,** 361 (1934).
56. K. AKITA and F. YOSHIDA, Phase-equilibria in methanol–ethyl acetate–water system, *J. Chem. Eng. Data* **8,** 484 (1963).
57. R.W. HANKINSON and D. THOMPSON, Equilibria and solubility data for the methanol–water–1-nitropropane ternary liquid system, *J. Chem. Eng. Data* **10,** 18 (1965).
58. G.V. JEFFREYS, Phase equilibria for system methylethylketone, cyclohexane and water, *J. Chem. Eng. Data* **8,** 320 (1963).

59. J.B.Claffey, C.O.Badgett, J.J.Skalcunera and G.W.Macpherson Phillips, Nicotine extraction from water with kerosene, *Ind. Eng. Chem.* **42,** 166 (1950).
60. K.S.Narasimhan, C.C.Reddy and K.S.Chari, Solubility and equilibrium data of phenol–water–n-butyl acetate system at 30 °C, *J. Chem. Eng. Data* **7,** 340 (1962).
61. K.S.Narasimhan, C.C. Reddy and K.S.Chari, Solubility and equilibrium data of phenol–water–isoamyl acetate and phenol–water–isobutyl ketone systems at 30 °C, *J. Chem. Eng. Data* **7,** 457 (1962).
62. M.R.Rao and C.V.Rao, Ternary liquid equilibria: water–fatty acid–solvent systems, *J. Appl. Chem.* **6,** 269 (1956).
63. R.J.Rao and C.V.Rao, Ternary liquid equilibria systems: n-propanol–water–esters, *J. Appl. Chem.* **9,** 69 (1959).
64. W.W.Chew and V.Orr, Mutual solubility in the ternary system n-propyl alcohol–water–hexamethyldisiloxane, *J. Chem. Eng. Data* **4,** 215 (1959).
65. K.W.Warner, Extraction of uranyl nitrate in a disc column, *Chem. Eng. Sci.* **3,** 161 (1954).
66. A.Orell, Phase equilibria and interfacial tension system, Ethylene glycol–acetic acid–ethyl acetate, *J. Chem. Eng. Data* **12,** 1 (1967).
67. M.R.Rao and C.V.Rao, Ternary liquid equilibria, *J. Sci. and Ind. Research (India)* **14B,** 204 (1955).
68. H.M.Trimble and G.E.Frazer, Solubility of ethylene glycol—some ternary systems, *Ind. Eng. Chem.* **21,** 1063 (1929).
69. K.Ishida, Solvent selectivity of liquid ammonia for pure hydrocarbons—tie line data and comparison of selectivity with other solvents, *Bull. Chem. Soc. Japan* **30,** 612 (1957).
70. T.G.Hunter and T.Brown, Distribution in hydrocarbon-solvent systems, *Ind. Eng. Chem.* **39,** 1343 (1947).
71. H.I.Weck and H.Hunt, Vapour–liquid equilibria in the ternary system benzene–cyclohexane–nitromethane and the three binaries, *Ind. Eng. Chem.* **46,** 2521 (1954).
72. G.M.Hartig, G.C.Hood and R.L.Maycock, Quaternary liquid systems with three liquid phases, *J. Phys. Chem.* **59,** 52 (1955).
73. T.E.Degaleesan and G.S.Laddha, Ternary systems: propylene glycol–ethanol–hydrocarbons, *J. Appl. Chem.* **12,** 111 (1962).
74. N.A.Kolosovskii and F.S.Kulihov, Distribution of fatty acids between glycerol and other organic solvents, *J. Gen. Chem. (U.S.S.R.)* **5**(66), 1037 (1935a).
75. H.W.Smith, The nature of secondary valence. IV. Partition coefficients in system glycerine: acetone, *J. Phys. Chem.* **25,** 721 (1921).
76. B.G.Kyle and T.M.Reed, Some ternary liquid systems containing fluorocarbons, *J. Chem. Eng. Data* **5,** 266 (1960).
77. M.G.Boobar, P.M.Kerschner, R.T.Struck, S.A.Herbert, N.L.Griver and C.R.Kurney, Styrene ethylbenzene–diethylene glycol system, ternary, saturation, equilibrium diagram, *Ind. Eng. Chem.* **43,** 2922 (1951).
78. P.L.Chueh and S.W.Briggs, Liquid–liquid phase equilibria of systems triolein–furfural–n-heptane and trilinolein–furfural–n-heptane, *J. Chem. Eng. Data* **9,** 207 (1964).
79. S.J.Bates and H.D.Kirschman, The vapour pressure and free energies of the hydrogen halides in aqueous solution; the free energy of formation of hydrogen chloride, *J. Am. Chem. Soc.* **41,** 1991 (1919).

80. J. LINDNER, The electrolytic dissociation of sulphurous acid, *Monatshefte für Chemie* **33,** 613 (1912).
81. K. T. KOONCE and R. KOBAYASHI, A method for determining the solubility of gases in relatively nonvolatile liquids—Solubility of methane in n-decane, *J. Chem. Eng. Data* **9,** 490 (1964).

APPENDIX 6a
TO FIT A STRAIGHT LINE TO DATA BY THE METHOD OF LEAST SQUARES— THE REGRESSION OF y ON x

Suppose we have a number of points (x, y) as shown in Fig. A6a.1 to which we wish to fit the straight line

$$Y = m_y x + c_y \qquad \text{A6a(1)}$$

where Y is the best estimate of y for a given value of x, m_y is the slope of the line and c_y its intercept on the y-axis. The deviation of y from Y is $(y - Y)$ and the sum of the squares of all the deviations is given by

$$D = \Sigma\,(y - (m_y x + c_y))^2$$

We wish to choose m_y and c_y so that D is a minimum which involves setting the partial derivatives of D with respect to m_y and c_y both equal zero. Now

$$\left(\frac{\partial D}{\partial c_y}\right)_{my} = -2\Sigma\,(y - (m_y x + c_y))$$

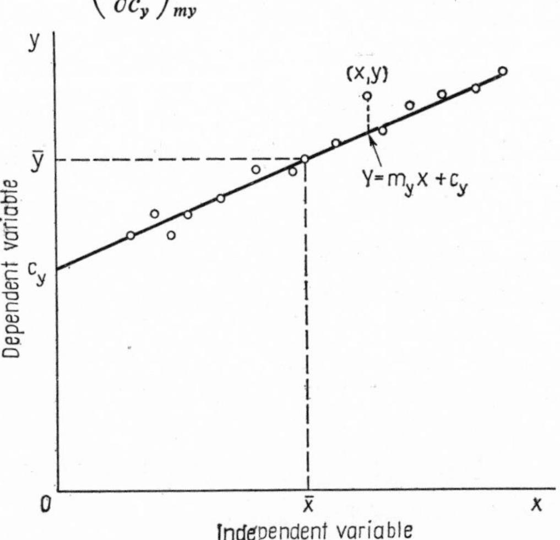

FIG. A6a.1. Illustration of method of least squares.

and

$$\left(\frac{\partial D}{\partial m_y}\right)_{c_y} = -2\Sigma\,(xy - x\,(m_y x + c_y))$$

so setting the partial derivatives equal to zero gives

$$\Sigma y = \Sigma\,(m_y x + c_y) = m_y \Sigma x + \Sigma c_y$$

and

$$\Sigma xy = \Sigma\,(m_y x^2 + c_y x) = m_y \Sigma x^2 + c_y \Sigma x$$

If there are p pairs of observations

$$\Sigma c_y = pc_y \colon \Sigma y = p\bar{y} \quad \text{and} \quad \Sigma x = p\bar{x}$$

so

$$c_y = \bar{y} - m_y \bar{x} \qquad \text{A6a(2)}$$

and

$$m_y = \frac{\Sigma x\,(y - \bar{y})}{\Sigma x\,(x - \bar{x})} \qquad \text{A6a(3)}$$

which may be re-expressed in the form

$$m_y = \frac{\Sigma\,(x - \bar{x})\,(y - \bar{y})}{\Sigma\,(x - \bar{x})^2} \qquad \text{A6a(4)}$$

Combining equations A6a(1) and (2) gives

$$Y - \bar{y} = m_y\,(x - \bar{x}) \qquad \text{A6a(5)}$$

which shows that the straight line passes through the mean values of x and y. The line is the regression of y upon x and m_y is known as the regression coefficient.

THE ERRORS IN m_y AND OTHER VARIABLES

An estimate of the variance in y is,

$$e_y^2 = \frac{\Sigma\,(y - (m_y x + c_y))^2}{p - 2} \qquad \text{A6a(6)}$$

as two degrees of freedom have been used up in estimating m_y and c_y. The variance of the mean value of y is,

$$e_{\bar{y}}^2 = e_y^2/p \qquad \text{A6a(7)}$$

and the variance in the mean value of x is zero as there is assumed to be no error in x, for we are predicting values of y from given values of x.

Because $\Sigma x\bar{y} = \dfrac{\Sigma x \Sigma y}{p} = \Sigma \bar{x} y$, equation A 6a(3) may be re-expressed in the form

$$m_y = \frac{\Sigma (x - \bar{x})\, y}{\Sigma (x - \bar{x})^2} \qquad \text{A 6a(8)}$$

$$= \frac{(x_1 - \bar{x})\, y_1 + (x_2 - \bar{x})\, y_2 + \ldots + (x_p - \bar{x})\, y_p}{\Sigma (x - \bar{x})^2} \qquad \text{A 6a(9)}$$

The variance in m_y is related to the variances in y by[10]

$$e_{m_y}^2 = \left(\frac{\partial m_y}{\partial y_1}\right)^2 e_{y_1}^2 + \left(\frac{\partial m_y}{\partial y_2}\right) e_{y_2}^2 + \ldots + \left(\frac{\partial m_y}{\partial y_p}\right)^2 e_{y_p}^2$$

For the ith pair of values x_i, y_i it follows from equation A 6a(9) that

$$\frac{\partial m_y}{\partial y} = \frac{x - \bar{x}}{\Sigma (x - \bar{x})^2}$$

and so

$$e_{m_y}^2 = \frac{(x_1 - \bar{x})^2 e_{y_1}^2 + (x_2 - \bar{x})^2 e_{y_2}^2 + \ldots + (x_p - \bar{x})^2 e_{y_p}^2}{(\Sigma (x - \bar{x})^2)^2}$$

If the error in each value of y is independent of the magnitude of y (as might be expected with concentration measurements) we may write

$$e_{y_1}^2 = e_{y_2}^2 = \ldots = e_{y_p}^2 = e_y^2$$

and so

$$e_{m_y}^2 = \frac{e_y^2\, \Sigma (x - \bar{x})^2}{(\Sigma (x - \bar{x})^2)^2}$$

$$= \frac{e_y^2}{\Sigma (x - \bar{x})^2} \qquad \text{A 6a(10)}$$

where an estimate of the variance in y, e_y^2 is given by equation A 6a(6).

The error in the value of y predicted by equation A 6a(5) is given by

$$e_{\hat{y}}^2 = e_{\bar{y}}^2 + (x - \bar{x})^2\, e_{m_y}^2 \qquad \text{A 6a(11)}$$

remembering there is no error in x or $\bar{x}$. It follows that the error in c_y (which is the value of y at $x = 0$) is given by

$$e_{c_y}^2 = e_{\bar{y}}^2 + \bar{x}^2 e_{m_y}^2 \qquad \text{A 6a(12)}$$

which could be obtained directly from equation A 6a(2). The variances $e_{\bar{y}}^2$ and $e_{m_y}^2$ are defined by equations A 6a(7) and (10) in terms of e_y^2 which is given by equation A 6a(6).

APPENDIX 6b
TO FIT A STRAIGHT LINE TO DATA BY THE METHOD OF LEAST SQUARES— THE REGRESSION OF x ON y

We wish to fit a straight line of the form

$$y = m_x X + c_x \qquad \text{A 6b(1)}$$

where X is the best estimate of x for a given value of y. Interchanging x and Y in equation A 6a(1) gives

$$X = m_y^* y + c_y^*$$

where m_y^* and c_y^* are the inverses of m_y and c_y discussed in Appendix 6a. Equation A 6b(1) may be written

$$X = (1/m_x)\, y - c_x/m_x$$

and so

$$m_y^* = 1/m_x \quad \text{and} \quad c_y^* = -c_x/m_x$$

The equations describing the regression of x on y may thus be obtained from the corresponding equations in Appendix 6a describing the regression of y on x, if x and y are interchanged, m_y is replaced by $1/m_x$ and c_y by $-c_x/m_x$. Thus

$$c_x = \bar{y} - m_x \bar{x} \qquad \text{A 6b(2)}$$

$$m_x = \frac{\Sigma y\,(y - \bar{y})}{\Sigma y\,(x - \bar{x})} \qquad \text{A 6b(3)}$$

$$= \frac{\Sigma\,(y - \bar{y})^2}{\Sigma\,(y - \bar{y})\,(x - \bar{x})} \qquad \text{A 6b(4)}$$

and

$$y - \bar{y} = m_x\,(X - \bar{x}) \qquad \text{A 6b(5)}$$

ERRORS IN m_x AND OTHER VARIABLES

We may write
$$e_x^2 = \frac{\Sigma\,(x - (1/m_x)\,(y - c_x))^2}{p - 2} \qquad \text{A6b(6)}$$

and
$$e_{\bar{x}}^2 = e_x^2/p \qquad \text{A6b(7)}$$

where
$$m_x = \frac{\Sigma\,(y - \bar{y})^2}{\Sigma\,(y - \bar{y})\,x} \qquad \text{A6b(8)}$$

But note that
$$1/m_x = \frac{(y_1 - \bar{y})\,x_1 + (y_2 - \bar{y})\,x_2 + \dots + (y_p - \bar{y})\,x_p}{\Sigma\,(y - \bar{y})^2} \qquad \text{A6b(9)}$$

and so
$$e_{(1/m_x)}^2 = \frac{e_x^2}{\Sigma\,(y - \bar{y})^2}$$

where
$$e_{(1/m_x)}^2 = \left(\frac{\partial\,(1/m_x)}{\partial m_x}\right)^2 e_{m_x}^2$$
$$= e_{m_x}^2/m_x^4$$

Thus
$$e_{m_x}^2 = \frac{m_x^4\, e_x^2}{\Sigma\,(y - \bar{y})^2} \qquad \text{A6b(10)}$$
$$e_x^2 = e_{\bar{x}}^2 + (y - \bar{y})^2\, e_{m_x}^2/m_x^4 \qquad \text{A6b(11)}$$

and
$$e_{c_x}^2 = m_x^2 e_{\bar{x}}^2 + \bar{x}^2 e_{m_x}^2 \qquad \text{A6b(12)}$$

APPENDIX 6c
TO FIT A STRAIGHT LINE THROUGH THE ORIGIN BY THE METHOD OF LEAST SQUARES— THE REGRESSION OF y ON x

Suppose we have p pairs of observations (x, y) to which we wish to fit the line
$$Y = M_y x \qquad \text{A6c(1)}$$

where Y is the best estimate of y for a given value of x. The sum of the squares of the deviations is
$$D = \Sigma\,(y - M_y x)^2$$

and so

$$\frac{\mathrm{d}D}{\mathrm{d}M_y} = -2\Sigma\,(y - M_y x)\,x$$

Setting the derivative equal to zero gives

$$M_y = \Sigma xy/\Sigma x^2 \qquad \text{A 6c(2)}$$

ERRORS IN M_y AND OTHER QUANTITIES

An estimate of the variance in y is

$$e_y^2 = \frac{\Sigma\,(y - M_y x)}{p - 1} \qquad \text{A 6c(3)}$$

as one degree of freedom has been exhausted in estimating M_y. Equation A 6c(2) may be written

$$M_y = (x_1 y_1 + x_2 y_2 + \cdots + x_p y_p)/\Sigma x^2$$

and as there is assumed to be no error in x

$$e_{M_y}^2 = (x_1^2 e_{y_1}^2 + x_2^2 e_{y_2}^2 + \cdots + x_p^2 e_{y_p}^2)/(\Sigma x^2)^2$$

If the variance in each of values of y is constant and given by equation A 6c(3)

$$e_{M_y}^2 = e_y^2 \Sigma x^2/(\Sigma x^2)^2 = e_y^2/\Sigma x^2 \qquad \text{A 6c(4)}$$

The error in a predicted value of y is given by

$$e_y^2 = x^2 e_{M_y}^2 \qquad \text{A 6c(5)}$$

The equations correspond to those in Appendix 6a if we set $\bar{x} = \bar{y} = 0$.

APPENDIX 6d
TO FIT A STRAIGHT LINE THROUGH THE ORIGIN BY THE METHOD OF LEAST SQUARES—THE REGRESSION OF x ON y

We wish to fit the line

$$y = M_x X \qquad \text{A 6d(1)}$$

where X is the best estimate of x for a given value of y. Interchanging x and Y in equation A 6c(1) gives

$$X = M_y^* y$$

where M_y^* is the inverse of M_y; equation A6d(1) may be written

$$X = (1/M_x)\, y$$

and so $M_y^* = 1/M_x$. The equations describing the regression of x on y may thus be obtained from the corresponding equations for the regression of y on x if x and y are interchanged and M_y is replaced by $1/M_x$ so that

$$M_x = \Sigma y^2/\Sigma xy \qquad \text{A6d(2)}$$

and

$$e_x^2 = \Sigma\,(x - (1/M_x)\, y)/(p - 1) \qquad \text{A6d(3)}$$

But note that

$$1/M_x = (y_1 x_1 + y_2 x_2 + \cdots + y_n x_n)/\Sigma y^2$$

and so

$$e_{(1/M_x)}^2 = e_x^2/\Sigma y^2$$

Now

$$e_{(1/M_x)}^2 = e_{M_x}^2/M_x^4$$

so that

$$e_{M_x}^2 = M_x^4 e_x^2/\Sigma y^2 \qquad \text{A6d(4)}$$

and

$$e_x^2 = y^2 e_{M_x}^2/M_x^4 \qquad \text{A6d(5)}$$

These equations correspond to those in Appendix 6b if we set $\bar{x} = \bar{y} = 0$.

APPENDIX 6e
REDUCTION OF FUNCTIONAL RELATIONSHIP TO THE TWO REGRESSIONS

The functional relationship and the regressions of y on x and x on y all pass through the mean values of x and y as illustrated in Fig. A6e.1. Because the regression of y on x assumes there is error in the values of y but not in the values of x the line is biased towards the mean value of y (which is the best single estimate of y). Similarly the regression of x on y assumes the x values are in error but not the y values and so the line is biased towards the mean value of x. The two regressions are the limiting values of the functional relationship as illustrated below.

The functional relationship is

$$y - \bar{y} = m(x - \bar{x}) \qquad \text{A6e(1)}$$

where

$$m = b + (b^2 + k^2)^{\frac{1}{2}} \qquad \text{A6e(2)}$$

$$b = \frac{\Sigma (y - \bar{y})^2 - k^2 \Sigma (x - \bar{x})^2}{2\Sigma (x - \bar{x})(y - \bar{y})} \qquad \text{A6e(3)}$$

and

$$k = e_y/e_x \qquad \text{A6e(4)}$$

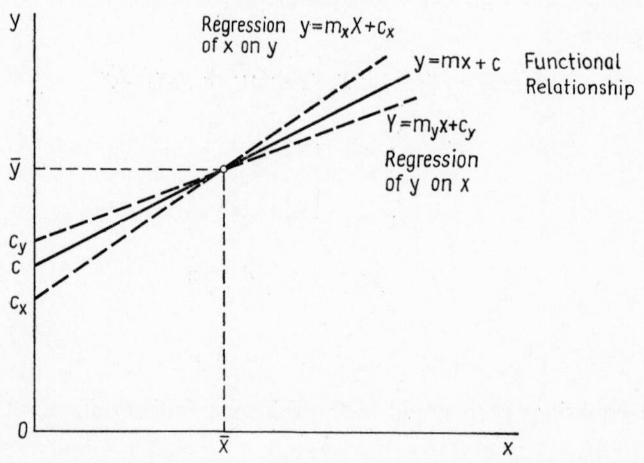

FIG. A6e.1. Illustration of two regression lines and functional relationship.

For the regression of y upon x

$$m_y = \frac{\Sigma (x - \bar{x})(y - \bar{y})}{\Sigma (x - \bar{x})^2} \qquad \text{A6e(5)}$$

and for the regression of x upon y

$$m_x = \frac{\Sigma (y - \bar{y})^2}{\Sigma (x - \bar{x})(y - \bar{y})} \qquad \text{A6e(6)}$$

so

$$b = \frac{1}{2}\left(m_x - \frac{k^2}{m_y}\right) \qquad \text{A6e(7)}$$

In the regression of x upon y, there is assumed to be no error in y so $e_y = 0$ and the ratio of the errors $k = 0$. Thus $b = m_x/2$ and $m = 2b = m_x$. The functional slope is thus identical with that resulting from the regression of x upon y when $e_y = 0$.

In the regression of y upon x, there is assumed to be no error in x so $e_x = 0$ and the ratio of the errors, k approaches infinity. Now

$$b^2 + k^2 = \frac{k^4}{4m_y^2}\left(1 + \frac{4m_y^2}{k^2}\left(1 - \frac{m_x}{2m_y}\right) + m_y^2 \frac{m_x^2}{k^4}\right)$$

and so neglecting the term in $1/k^4$

$$(b^2 + k^2)^{\frac{1}{2}} = \frac{k^2}{2m_y}\left(1 + \frac{2m_y^2}{k^2}\left(1 - \frac{m_x}{2m_y}\right)\right)$$

The functional slope thus becomes

$$\begin{aligned} m &= b + (b^2 + k^2)^{1/2} \\ &= \frac{m_x}{2} - \frac{k^2}{2m_y} + \frac{k^2}{2m_y} + m_y - \frac{m_x}{2} \\ &= m_y \end{aligned}$$

which is the slope resulting from the regression of y upon x when $e_x = 0$.

The variance in the functional slope is

$$e_m^2 = \frac{(k^2 + m^2)^2\, e_y^2/k^2}{k^2\Sigma(x - \bar{x})^2 + m\Sigma(x - \bar{x})(y - \bar{y})} \qquad \text{A6e(8)}$$

or

$$e_m^2 = \frac{(k^2 + m^2)^2\, e_x^2}{k^2\Sigma(x - \bar{x})^2 + m\Sigma(x - \bar{x})(y - \bar{y})} \qquad \text{A6e(9)}$$

where the variances in y and x may be estimated from

$$e_y^2 = \frac{\Sigma(y - \bar{y})^2 - m\Sigma(x - \bar{x})(y - \bar{y})}{p - 2} \qquad \text{A6e(10)}$$

and

$$e_x^2 = \frac{\Sigma(x - \bar{x})^2 - (1/m)\,\Sigma(x - \bar{x})(y - \bar{y})}{p - 2} \qquad \text{A6e(11)}$$

For the regression of x upon y the variance in the slope is given by equation A 6b(10) as

$$e_{m_x}^2 = \frac{m_x^4 e_x^2}{\Sigma (y - \bar{y})^2} \qquad \text{A 6e(12)}$$

where the variance in x given by equation A 6b(6) may be written

$$e_x^2 = \frac{\Sigma (x - \bar{x})^2 - (1/m_x) \Sigma (x - \bar{x})(y - \bar{y})}{p - 2} \qquad \text{A 6e(13)}$$

Similarly for the regression of y upon x the variance in the slope is given by equation A 6a(10) as

$$e_{m_y}^2 = \frac{e_y^2}{\Sigma (x - \bar{x})^2} \qquad \text{A 6e(14)}$$

where the variance in y given by equation A 6a(6) may be written

$$e_y^2 = \frac{\Sigma (y - \bar{y})^2 - m_y \Sigma (x - \bar{x})(y - \bar{y})}{p - 2} \qquad \text{A 6e(15)}$$

In the regression of x on y, $e_y^2 = 0$ and so $k = 0$; putting $m = m_x$ in equation A 6e(9) and using the definition of m_x in equation A 6e(6) gives

$$e_m^2 = \frac{m_x^4 e_x^2}{\Sigma (y - \bar{y})^2}$$

where the variances in x defined by equations A 6e(11) and (13) are identical. The value of e_m^2 thus equals the value of $e_{m_x}^2$ defined by equation A 6e(12) and the variance of the functional slope thus reduces to that of the regression of x upon y when $e_y = 0$.

In the regression of y on x, $e_x^2 = 0$ so k approaches infinity, and equation A 6e(9) becomes

$$e_m^2 = \frac{e_y^2}{\Sigma (x - \bar{x})^2}$$

Putting $m = m_y$ shows that the variances in y defined by equations A 6e(10) and (15) are identical. The value of e_m^2 thus equals the value of $e_{m_y}^2$ defined by equation A 6e(14) and the variance in the functional slope thus reduces to that of the regression of y on x when $e_x = 0$.

APPENDIX 6f
COMPUTER PROGRAM 6.1. FUNCTIONAL RELATIONSHIPS AND LEAST SQUARES REGRESSIONS FOR EQUILIBRIUM DATA

It is generally easier to calculate sums like Σx^2 rather than $\Sigma (x - \bar{x})^2$ and so the equations were first re-expressed in this form before being used for computation. For example, equations A6a(4) and A6b(4) become

$$m_y = \frac{\Sigma xy - p\bar{x}\bar{y}}{\Sigma x^2 - p\bar{x}^2}$$

and

$$m_x = \frac{\Sigma y^2 - p\bar{y}^2}{\Sigma xy - p\bar{x}\bar{y}}$$

In addition it follows from equations A6a(10) and A6b(10) that the fractional errors in these slopes are equal and given by

$$\frac{e_{m_y}^2}{m_y^2} = \frac{e_{m_x}^2}{m_x^2} = \frac{1}{p-2}\left(\frac{m_x}{m_y} - 1\right)$$

Only the percentage error in the slope for one of the regressions need thus be determined. This also applies to lines through the origin if m is replaced by M, and $p - 2$ by $p - 1$.

A typical computer program 3/24A, written in Atlas Autocode, is presented below together with a sample of the input data and the corresponding results obtained from a KDF9 computer. The program reads in a set of (x, y) data for ascending order of concentration. For lines through the origin it calculates the functional slope and its error, and the slopes of the two regressions and their fractional error. The program then reverses the data so it becomes in decreasing order of concentration and calculates the slopes and errors of the lines through the mean values of x and y for the different numbers of points considered. Another set of data is then read in and processed in the same way. The program is intended to be self-explanatory rather than efficient.

```
***A

JOB
UNOTT. CHE/HARTLAND 3/24A
EXECUTION 2 MINUTES
OUTPUT 0 LINEPRINTER 1500 LINES
STORE 30/30 BLOCKS
COMPILER AA

begin
array x,y,x',y'(0:100)
real X,Y,xy,xx,yy,My,Mx,M,b,eey,eex,ee,ey,ex,e,my,mx,mxy,m,xav,yav,k
integer i,j,l,n
caption ╪least$squares$regression╪gas-liquid$equilibria╪╪

0: i=0; k=1
1: read(X); ->2 if X=999999; read(Y); i=i+1; x(i)=X; y(i)=Y; ->1
2: ->5 if i=0
caption ╪╪╪$$n$$$$$$$x$$$$$$$$$$$$$y$$$$$$$$$$$$My$$$$$$$$$$Mx$$$$$$$$$Mxy
caption $$$$$$$$$M$$$$$$$$peMy$$$$$peM╪
xy=0; xx=0; yy=0
cycle n=1,1,i; comment fitting line of the form y=Mx
xy=xy+x(n)*y(n); xx=xx+x(n)*x(n); yy=yy+y(n)*y(n)
My=xy/xx; print(n,2,0); print(x(n),4,6); print(y(n),4,6); print(My,4,6)
Mx=yy/xy; print(Mx,4,6); print((My+Mx)/2,4,6); b=(yy-k²*xx)/(2xy)
M=b+sqrt(k²+b²); print(M,4,6), ->3 if n=1
eey=(yy/xx-My²)/(n-1); ey=sqrt(mod(eey)); print(100ey/My,3,3)
ee=((k²+M²)²*(yy-M*xy))/(k²*(n-1)*(k²*xx+M*xy)); e=sqrt(mod(ee))
print(100e/M,3,3)
3: newline; repeat; newline; comment inversion of data
cycle =1,1,i; j=i-l+1; x'(j)=x(l); y'(j)=y(l); repeat
comment fitting line of the form (y-yav)=m(x-xav)
caption $$n$$$$$$$xav$$$$$$$$$yav$$$$$$$$$$my$$$$$$$$$mx$$$$$$$$$$mxy$$$$
caption $$$$$$m$$$$$$pemy$$pem$$$$$$eyav$$$$$$$exav$$$$$$e(xy)av╪
xy=0; xx=0; yy=0; X=0; Y=0
cycle n=1,1,l; print(n,2,0); xy=xy+x'(n)*y'(n); xx=xx+x'(n)*x'(n)
yy=yy+y'(n)*y'(n); X=X+x'(n); Y=Y+y'(n); xav=X/n; yav=Y/n
->4 if n<2.5
my=(xy-yav*X)/(xx-xav*X); mx=(yy-yav*Y)/(xy-xav*Y)
mxy=(my+mx)/2
b=(yy-k²*xx+k²*xav*Y)/(2xy-2xav*Y); m=b+sqrt((k²+b²)); print(xav,4,6)
print(yav,4,6); print(my,4,6); print(mx,4,6); print(mxy,4,6); print(m,4,6)
eey=((yy-n*yav²)/(xx-n*xav²)-my²)/(n-2); ey=sqrt(mod(eey))
print(100ey/my,2,2)
ee=(k²+m²)²*(yy-yav*Y-m*xy+m*xav*Y)
ee=ee/(k²(n-2)*(k²*xx-k²*X*xav+m*xy-m*yav*X))
```

```
e=sqrt(mod(ee)); print(100e/m,2,2); eey=eey*(xx-n*xav²)/n
ey=sqrt(mod(eey)); print(ey,3,6)
eex=(mx²*(xx-X*xav)/(yy-Y*yav)-1)*(yy-Y*yav)/(mx²*(n²-2n))
ex=sqrt(mod(eex)); print(ex,3,6); ee=(yy-Y*yav-m*xy+m*xav*Y)/(n²-2n)
e=sqrt(mod(ee)); print(e,3,6)
4: newline
repeat; ->0
5: stop
end of program
```

Input Data

```
544.6    300
711.7    500
866.2    700
1004.7   900
1145.7  1100
1286.9  1300
1428.0  1500   999 999

0.997    900
1.094   1000
1.192   1100
1.288   1200
1.388   1300
1.488   1400
1.588   1500
1.688   1600
1.778   1700
1.847   1800
1.887   1865   999 999
```

***Z

Output Data

24/08/67 12.23.18
EDINBURGH UNIVERSITY ATLAS AUTOCODE 12/05/66

UNOTT. CHE/HARTLAND 3/24A
0 BEGIN
39 END OF PROGRAM

PROGRAM (+PERM) OCCUPIES 2880 WORDS
PROGRAM DUMPED
COMPILING TIME 2 MIN 18 SEC / 12 SEC

LEAST SQUARES REGRESSION
GAS-LIQUID EQUILIBRIA

N	X	Y	MY	MX	MXY	M	PEMY	PEM
1	544.600000	300.000000	0.550863	0.550863	0.550863	0.550863		
2	711.700000	500.000000	0.646527	0.654816	0.650672	0.648977	11.323	11.292
3	866.200000	700.000000	0.724581	0.737404	0.730992	0.729013	9.407	9.369
4	1004.700000	900.000000	0.792015	0.807961	0.799988	0.798191	8.192	8.153
5	1145.700000	1100.000000	0.848950	0.866243	0.857596	0.856229	7.136	7.101
6	1286.900000	1300.000000	0.897221	0.914762	0.905991	0.905082	6.253	6.223
7	1428.000000	1500.000000	0.938485	0.955660	0.947072	0.946565	5.523	5.498

N	XAV	YAV	MY	MX	MXY	M	PEMY	PEM	EYAV	EXAV	E(XY)AV
1											
2											
3	1286.866667	1300.000000	1.416932	1.416932	1.416932	1.416932	0.02	0.02	0.033405	0.023575	0.019401
4	1216.325000	1200.000000	1.417334	1.417334	1.417334	1.417334	0.02	0.02	0.041959	0.029604	0.024159
5	1146.300000	1100.000000	1.422657	1.422677	1.422667	1.422671	0.22	0.22	0.616419	0.433283	0.354473
6	1073.866667	1000.000000	1.404003	1.404325	1.404164	1.404217	0.76	0.76	2.588646	1.843549	1.501699
7	998.257143	900.000000	1.371155	1.372515	1.371835	1.372043	1.41	1.41	5.632348	4.105704	3.318154

N	X	Y	MY	MX	MXY	M	PEMY	PEM
1	0.997000	900.000000	902.708124	902.708124	902.708124	902.708124		
2	1.094000	1000.000000	908.918705	908.953950	908.936327	908.953950	0.623	0.619
3	1.192000	1100.000000	914.387067	914.458743	914.422905	914.458743	0.626	0.625
4	1.288000	1200.000000	919.829099	919.948021	919.888560	919.948021	0.656	0.659
5	1.388000	1300.000000	924.318176	924.464487	924.391331	924.464487	0.629	0.631
6	1.488000	1400.000000	928.209912	928.374372	928.292142	928.374372	0.595	0.595
7	1.588000	1500.000000	931.670226	931.847414	931.758820	931.847414	0.563	0.562
8	1.688000	1600.000000	934.792246	934.978471	934.885358	934.978471	0.533	0.532
9	1.778000	1700.000000	938.551570	938.774783	938.663176	938.774783	0.545	0.546
10	1.847000	1800.000000	944.302730	944.673381	944.488055	944.673381	0.660	0.660
11	1.887000	1865.000000	950.596348	951.161824	950.879086	951.161823	0.771	0.772

N	XAV	YAV	MY	MX	MXY	M	PEMY	PEM	EYAV	EXAV	E(XY)AV
1											
2											
3	1.837333	1788.333333	1506.687858	1508.095323	1507.391591	1508.095323	3.06	0.00	2.073220	0.001375	0.000000
4	1.800000	1741.250000	1326.624572	1335.936208	1331.280390	1335.936202	5.92	6.36	5.934260	0.004458	0.004784
5	1.757600	1693.000000	1211.190378	1222.126990	1216.658684	1222.126983	5.49	5.29	7.203562	0.005921	0.005705
6	1.712667	1644.166667	1148.049968	1157.100451	1152.575210	1157.100444	4.44	4.40	7.187666	0.006236	0.006176
7	1.666286	1595.000000	1110.136466	1117.173236	1113.654851	1117.173230	3.56	3.59	6.841020	0.006143	0.006194
8	1.619000	1545.625000	1085.477794	1090.921751	1088.199772	1090.921746	2.89	2.92	6.420974	0.005901	0.005967
9	1.571556	1496.111111	1071.826544	1075.901112	1073.863828	1075.901109	2.33	2.30	5.869423	0.005466	0.005391
10	1.523800	1446.500000	1062.189270	1065.310744	1063.750007	1065.310741	1.92	1.90	5.394754	0.005071	0.005018
11	1.475909	1396.818182	1055.635961	1058.059877	1056.847919	1058.059875	1.60	1.60	4.966927	0.004700	0.004711

APPENDIX 6g
LIST OF SYSTEMS INVESTIGATED

This table lists the systems investigated divided into Aqueous–Organic systems (AO-1 to 99), Organic–Organic systems (OO-1 to 47) and Liquid–Gas systems (LG-1 to 39) arranged alphabetically according to the solute.

The concentration units cm^3/cm^3 refer to the cubic centimetres of gas (reduced to S.T.P.) dissolved in 1 cubic centimetre of liquid solvent.

Aqueous–Organic

System number	Solute	Solvent (y)	Temperature	Concentration units	Reference
AO-1	Acetaldehyde	Vinyl acetate	20°C	wt. %	33
AO-2	Acetic acid	1-Butanol	26·7°C	wt. %	34
AO-3		Cyclohexanol	26·7°C	wt. %	34
AO-4		Ethyl acetate	15°C	g/100 g nonsolute	35
AO-5		Ethyl butyrate	28°C	wt. %	17
AO-6		Furfural	25°C	wt. %	36
AO-7		Furfural	35°C	wt. %	36
AO-8		Heptadecanol	25°C	wt. %	37
AO-9		Isophorone	24°C	wt. %	16
AO-10		Methyl cyclohexanone	23°C	wt. %	16
AO-11		Methyl ethyl ketone	26·7°C	wt. %	34
AO-12		Methyl isobutyl carbinol	30°C	lb moles/ft^3	38
AO-13		Methyl isobutyl ketone	25°C	wt. %	18
AO-14		Methyl isobutyl ketone	28°C	wt. %	39
AO-15		Nitromethane	26·7°C	wt. %	34
AO-16		Octyl acetate	23°C	wt. %	16
AO-17		*o*-Xylene	29°C	wt. %	39

AO-18	Acetone	Benzene	15°C	wt. fraction	26
AO-19		Benzene	30°C	wt. fraction	26
AO-20		Benzene	45°C	wt. fraction	26
AO-21		Tetrachloro-ethane	23°C	wt. %	16
AO-22		Vinyl-acetate	20°C	wt. %	33
AO-23		*o*-Xylene	25°C	wt. %	16
AO-24		*o*-Xylene	30°C	wt. %	39
AO-25	Ammonia	Propane	20°C	wt. fraction	40
AO-26		Propylene	20°C	wt. fraction	40
AO-27	Benzoic acid	Methyl isobutyl ketone	80°F	lb moles/ft^3	41
AO-28		Toluene	59·9°F	lb moles/ft^3	42
AO-29		Toluene	83·8°F	lb moles/ft^3	42
AO-30		Toluene	30°C	lb moles/ft^3	38
AO-31	1-Butanol	Methyl 1-butyl ketone	100°F	wt. %	43
AO-32	n-Butyric acid	Ethyl butyrate	28°C	wt. %	17
AO-33		Furfural	25°C	wt. %	44
AO-34		Furfural	35°C	wt. %	44
AO-35		1-Hexanol	30°C	wt. %	45
AO-36		Methyl isobutyl carbinol	30°C	wt. %	46
AO-37		Methyl isobutyl ketone	25°C	g moles/l	18
AO-38	n-Caproic acid	Methyl isobutyl ketone	20°C	g moles/l	18
AO-39	Diethylamine	Methyl isobutyl ketone	20°C	g moles/l	19
AO-40		Toluene	20°C	g moles/l	19
AO-41	Ethanol	Ethyl acetate	0°C	wt. %	47
AO-42		Ethyl acetate	20°C	wt. %	47
AO-43		Ethyl ether	25°C	wt. %	48
AO-44		Heptadecanol	25°C	wt. %	8
AO-45	Ethyl acetate	1-Butyraldehyde	100°F	wt. %	43
AO-46	Formic acid	Methyl isobutyl carbinol	30°C	wt. %	46
AO-47		Methyl isobutyl ketone	25°C	g moles/l	18

Aqueous–Organic (cont.)

System number	Solute	Solvent (y)	Temperature	Concentration units	Reference
AO-48		Methyl isobutyl ketone	25°C	g equiv/l	49
AO-49	Isovaleric acid	Benzene	25°C	g moles/l	50
AO-50		Chloroform	25°C	g moles/l	50
AO-51		Isobutanol	25°C	g moles/l	51
AO-52		Toluene	25°C	g moles/l	50
AO-53		*o*-Xylene	25°C	g moles/l	52
AO-54	Lactic acid	1-Butanol	25°C	wt. %	20
AO-55		Isoamyl alcohol	25°C	g/100 ml	53
AO-56	Malonic acid	Ethyl ether	25°C	m moles/l	21
AO-57	Methanol	n-Butyl acetate	30°C	wt. %	54
AO-58		Cyclohexane	24·8°C	wt. %	55
AO-59		Ethyl acetate	70°C	wt. fraction	56
AO-60		Ethyl acetate	70°C	mole fraction	56
AO-61	Methanol	Ethyl butyrate	30°C	wt. %	54
AO-62		Ethyl propionate	30°C	wt. %	54
AO-63		1-Nitropropane	25°C	wt. %	57
AO-64		Amyl acetate	30°C	wt. %	54
AO-65	Methyl ethyl ketone	Cyclohexane	25°C	wt. fraction	58
AO-66		n-Hexane	100°F	wt. %	43
AO-67	Nicotine	Benzene	25°C	g/l	22
AO-68		Carbon tetrachloride	25°C	g/l	22
AO-69		Chloroform	25°C	g/l	22
AO-70		Kerosene	20°C	g/l	59
AO-71		Kerosene	35°C	g/l	59
AO-72		Kerosene	50°C	g/l	59

AO-73	Oxalic acid	n-Amyl alcohol	room temperature	g/l	23
AO-74	Phenol	n-Butyl acetate	30°C	wt. fraction	60
AO-75		Isoamyl acetate	30°C	wt. fraction	61
AO-76		Methyl isobutyl ketone	30°C	wt. fraction	61
AO-77	Propionic acid	Cyclohexane	31°C	wt. %	62
AO-78		Cyclohexane	31°C	wt. %	62
AO-79		Furfural	25°C	wt. %	36
AO-80		Furfural	35°C	wt. %	36
AO-81		Hexane	31°C	wt. %	62
AO-82		1-Hexanol	30°C	wt. %	45
AO-83		Methyl butyrate	30°C	wt. %	24
AO-84		Methyl isobutyl carbinol	30°C	wt. %	46
AO-85		Olive oil	25°C	g moles/l	25
AO-86		Tetrachloroethane	31°C	wt. %	62
AO-87		Toluene	31°C	wt. %	62
AO-88	n-Propanol	Amyl acetate	30°C	wt. %	63
AO-89		n-Butyl acetate	30°C	wt. %	63
AO-90		Ethyl butyrate	30°C	wt. %	63
AO-91		Ethyl propionate	30°C	wt. %	63
AO-92		Hexamethyl-disiloxane	25°C	wt. %	64
AO-93		n-Propyl acetate	30°C	wt. %	63
AO-94	Succinic acid	n-Amyl alcohol	room temperature	g/l	23
AO-95	Sulphuric acid	Methyl isobutyl ketone	25°C	g equiv/l	49
AO-96	Triethylamine	Methyl isobutyl ketone	20°C	g moles/l	19
AO-97	Uranyl nitrate	Dibutyl carbitol	20°C	lb/lb solution	65
AO-98		Methyl isobutyl ketone	20°C	lb/lb solution	65
AO-99	n-Valeric acid	Methyl isobutyl ketone	25°C	g moles/l	18

Organic–Organic

System number	Solute	Solvent (x)	Solvent (y)	Temperature	Concentration units	Reference
OO-1	Acetic acid	Ethylene glycol	Ethyl acetate	25°C	wt. fraction	66
OO-2	Acetone	Ethylene glycol	Amyl acetate	31°C	wt. %	67
OO-3			Benzene	27°C	vol. %	68
OO-4			Bromobenzene	25°C	vol. %	68
OO-5			n-Butyl acetate	31°C	wt. %	67
OO-6			Chlorobenzene	23°C	vol. %	68
OO-7			Cyclohexane	27°C	wt. %	67
OO-8			Ethyl acetate	31°C	wt. %	67
OO-9			Ethyl butyrate	31°C	wt. %	67
OO-10			Ethyl propionate	31°C	wt. %	67
OO-11			Nitrobenzene	22°C	vol. %	68
OO-12			Toluene	0°C	mole fraction	2
OO-13				24°C	mole fraction	2
OO-14				27°C	vol. %	68
OO-15			*o*-Xylene	25°C	vol. %	68
OO-16	Benzene	Ammonia	Cyclohexane	0°C	wt. %	69
OO-17				20°C	wt. %	69
OO-18			n-Hexane	20°C	wt. %	69
OO-19		Cetane	Aniline	25°C	wt. %	70
OO-20		Cyclohexane	Nitromethane	25°C	wt. %	71
OO-21		n-Heptane	Sulfolane	25°C	wt. %	72
OO-22	Cyclohexane	Cetane	Aniline	25°C	wt. %	70
OO-23		n-Heptane	Aniline	25°C	wt. %	70
OO-24	Diphenyl-hexane	Docosane	Furfural	45°C	wt. %	26

OO-25	Ethanol	Propylene glycol	Benzene	30°C	wt. %	73
OO-26			Cyclohexane	30°C	wt. %	73
OO-27			n-Hexane	30°C	wt. %	73
OO-28	n-Heptane	Cetane	Aniline	25°C	wt. %	70
OO-29	Isovaleric acid	Glycerol	Chloroform	25°C	g moles/l	74
OO-30			Nitrobenzene	25°C	g moles/l	74
OO-31			Toluene	25°C	g moles/l	74
OO-32	Malonic acid	Glycerol	Acetone	25°C	mg moles/l	75
OO-33	Perfluoro-cyclic oxide	Carbon tetra-chloride	Perfluoroheptane	30°C	wt. fraction	76
OO-34		n-Heptane	Perfluoroheptane	30°C	wt. fraction	76
OO-35	Propionic acid	Glycerol	Chloroform	25°C	g moles/l	74
OO-36	Propylene	Propane	Ammonia	0°C	wt. fraction	40
OO-37	Styrene	Ammonia	Ethylbenzene	−15·5°C	wt. %	69
OO-38				0°C	wt. %	69
OO-39		Diethylene glycol	Ethylbenzene	25°C	wt. %	77
OO-40	Toluene	Methyl-cyclo-hexane	Methyl perfluoro-octanoate	10°C	wt. fraction	76
OO-41				25°C	wt. fraction	76
OO-42	Trilinolein	Furfural	n-Heptane	30°C	wt. fraction	78
OO-43				50°C	wt. fraction	78
OO-44				70°C	wt. fraction	78
OO-45	Triolein	Furfural	n-Heptane	30°C	wt. fraction	78
OO-46				50°C	wt. fraction	78
OO-47				70°C	wt. fraction	78

Liquid–Gas

System number	Gas (y)	Liquid (x)	Temperature	Concentration units		Reference
				y-phase	x-phase	
LG-1	Ammonia	Water	0°C	mm Hg partial	g/g	27
LG-2					cm^3/cm^3	
LG-3			10°C	mm Hg total	g/g	27
LG-4					cm^3/cm^3	
LG-5			20°C	mm Hg partial	g/g	27
LG-6					cm^3/cm^3	
LG-7				mm Hg total	g/g	27
LG-8					cm^3/cm^3	
LG-9			30°C	mm Hg total	g/g	27
LG-10					cm^3/cm^3	
LG-11			40°C	mm Hg partial	g/g	27
LG-12					cm^3/cm^3	
LG-13	Carbon dioxide	Water	20°C	atm. total	cm^3/cm^3	28
LG-14			35°C			28
LG-15						
LG-16			60°C			28
LG-17						
LG-18			100°C			28
LG-19						
LG-20	Hydrogen bromide	Water	25°C	mm Hg partial	g/g	79
LG-21					cm^3/cm^3	
LG-22	Hydrogen chloride	Water	25°C	mm Hg partial	g/g	79
LG-23					cm^3/cm^3	
LG-24	Hydrogen iodide	Water	25°C	mm Hg partial	g/g	79
					cm^3/cm^3	

LG-25						
LG-26	Propylene	Water	100°F	psia total	mole %	29
LG-27	Sulphur dioxide	Water	25°C	mm Hg partial	mg/cm³ solution	80
LG-28			50°C			80
LG-29	Methane	n-Decane	−20°F	mm Hg partial	mole fraction	81
LG-30			0°F			81
LG-31			20°F			81
LG-32			40°F			81
LG-33	Dimethyl ether	Acetone	25°C	mm Hg partial	mole fraction	30
LG-34		Benzene				30
LG-35		Carbon tetrachloride				30
LG-36		Chlorobenzene				30
LG-37		Methyl acetate				30
LG-38	Sulphur dioxide	Chloroform	0°C	mm Hg partial	g/l solution	80
LG-39			25°C			

APPENDIX 6h
SLOPES AND ERRORS OF LINES THROUGH ORIGIN

Only systems having a minimum error in the slope of less than 4% are shown. Three concentration ranges are quoted for each system, these being from top to bottom;

(i) the full concentration range,
(ii) the range over which the error in the slope is a minimum,
(iii) the widest range over which the error in the slope is less than about 2%.

Aqueous–Organic Systems

System number	Upper concentration in *x*-phase	Slope *M*	% error in *M*	System number	Upper concentration in *x*-phase	Slope *M*	% error in *M*
	10·0	0·765	2·73		31·0	0·979	0·33
AO-1	10·0	0·765	2·73	AO-10	31·0	0·979	0·33
	10·0	0·765	2·73		31·0	0·979	0·33
	12·0	1·294	1·45		3·08	1·354	1·95
AO-2	6·35	1·358	1·10	AO-11	1·41	1·252	1·49
	12·0	1·294	1·45		3·08	1·354	1·95
	22·8	1·063	3·18		0·02539	0·875	2·00
AO-3	8·35	1·251	1·53	AO-12	0·01148	0·938	0·87
	11·3	1·190	1·84		0·01738	0·911	1·47
	28·5	0·892	1·49		2·870	0·619	3·04
AO-4	1·43	1·175	0·06	AO-13	0·456	0·492	0·78
	28·5	0·892	1·49		1·240	0·532	1·91
	40·1	0·521	2·10		22·40	0·804	1·71
AO-5	40·1	0·521	2·10	AO-14	1·398	0·617	0·33
	40·1	0·521	2·10		19·88	0·789	1·68
	15·4	0·874	3·19		20·60	0·503	6·30
AO-6	7·0	0·750	1·89	AO-15	16·30	0·450	1·76
	7·0	0·750	1·89		16·30	0·450	1·76
	12·5	0·854	2·64		60·2	0·319	6·41
AO-7	10·2	0·819	2·20	AO-16	14·1	0·178	0·74
	10·2	0·819	2·20		21·2	0·188	2·80
	82·0	0·406	13·24		0·500	1·342	2·74
AO-8	48·4	0·249	2·33	AO-18	0·500	1·342	2·74
	48·4	0·249	2·33		0·500	1·342	2·74
	26·6	1·010	1·98		0·500	1·409	3·79
AO-9	6·8	0·863	0·61	AO-19	0·400	1·503	2·16
	26·6	1·010	1·98		0·400	1·503	2·16

Aqueous–Organic Systems (cont.)

System number	Upper concentration in x-phase	Slope M	% error in M	System number	Upper concentration in x-phase	Slope M	% error in M
	0·400	1·605	3·10		0·01125	79·406	0·95
AO-20	0·300	1·699	2·00	AO-38	0·01125	79·406	0·95
	0·300	1·699	2·00		0·01125	79·406	0·95
	10·0	1·294	1·18		0·1842	0·663	0·85
AO-22	5·0	1·221	0·08	AO-39	0·1842	0·663	0·85
	9·0	1·273	1·00		0·1842	0·663	0·85
	34·3	0·987	7·10		0·590	0·557	0·61
AO-23	22·3	0·738	2·92	AO-40	0·590	0·557	0·61
	22·3	0·738	2·92		0·590	0·557	0·61
	22·0	0·854	2·89		19·5	0·852	5·00
AO-24	1·8	0·658	0·25	AO-42	13·7	0·733	3·95
	9·0	0·742	1·94		13·7	0·733	3·95
	0·764	0·167	17·68		63·9	0·399	6·44
AO-26	0·279	0·0425	1·39	AO-44	2·17	0·292	2·58
	0·279	0·0425	1·39		21·7	0·292	2·58
	0·00111	72·277	1·30		54·6	0·382	3·38
AO-27	0·00111	72·277	1·30	AO-46	54.6	0·382	3·38
	0·00111	72·277	1·30		54·6	0·382	3·38
	6·40	10·482	2·86		2·768	0·445	0·72
AO-31	6·40	10·482	2·86	AO-47	0·464	0·421	0·37
	6·40	10·482	2·86		2·678	0·445	0·72
	6·5	6·635	0·73		8·846	0·530	4·72
AO-32	6·5	6·635	0·73	AO-48	1·631	0·424	0·46
	6·5	6·635	0·73		7·369	0·472	1·86
	11·0	4·032	4·10		0·379	14·988	2·34
AO-33	2·0	4·436	0·33	AO-51	0·004	12·933	1·33
	6·1	4·572	1·04		0·308	14·402	2·31
	9·4	4·159	4·55		0·452	17·165	3·63
AO-34	4·0	4·619	0·29	AO-53	0·452	17·165	3·63
	5·8	4·665	0·53		0·452	17·165	3·63
	8·3	7·150	4·52		12·5	0·928	0·75
AO-35	1·7	9·636	2·02	AO-54	10·3	0·919	0·56
	1·7	9·636	2·02		12·5	0·928	0·75
	0·1559	6·806	1·53		22·38	0·526	3·10
AO-37	0·0803	6·413	0·89	AO-55	5·8	0·445	0·76
	0·1559	6·806	1·53		16·24	0·489	1·93

Aqueous–Organic Systems (cont.)

System number	Upper concentration in x-phase	Slope M	% error in M	System number	Upper concentration in x-phase	Slope M	% error in M
	94·6	0·148	0·59		155·0	1·268	7·56
AO-56	3·8	0·135	0·19	AO-72	8·6	2·459	0·40
	94·6	0·148	0·59		30·7	2·245	1·28
	38·6	0·312	9·62		81·6	0·362	1·09
AO-57	7·6	0·157	1·49	AO-73	81·6	0·362	1·09
	7·6	0·157	1·49		81·6	0·362	1·09
	41·8	0·0160	2·98		0·0880	11·441	8·27
AO-58	15·1	0·0165	1·14	AO-75	0·0050	33·811	1·78
	15·1	0·0165	1·14		0·0050	33·811	1·78
	0·154	1·230	3·52		0·0865	13·741	10·69
AO-60	0·123	1·302	1·17	AO-76	0·0067	40·004	1·16
	0·123	1·302	1·17		0·0080	40·505	1·17
	36·0	0·826	13·54		72·7	0·510	9·00
AO-62	15·5	0·153	1·79	AO-77	57·4	0·373	3·30
	15·5	0·153	1·79		57·4	0·373	3·30
	46·6	0·429	9·04		61·0	0·573	2·16
AO-64	12·6	0·160	0·81	AO-78	54·3	0·557	2·08
	12·6	0·160	0·81		54·3	0·557	2·08
	0·212	2·673	3·03		12·8	1·713	3·26
AO-65	0·212	2·673	3·03	AO-79	6·8	1·860	0·41
	0·212	2·673	3·03		9·4	1·819	1·16
	1·87	10·317	0·29		11·4	1·731	3·10
AO-67	1·65	10·276	0·23	AO-80	4·1	1·884	0·58
	1·87	10·317	0·29		8·2	1·833	0·99
	1·97	9·551	0·73		70·5	0·431	3·75
AO-68	1·69	9·641	0·67	AO-81	58·0	0·387	3·29
	1·97	9·551	0·73		58·0	0·387	3·29
	0·38	8·738	1·46		21·2	2·159	6·93
AO-69	0·24	8·688	1·25	AO-82	4·8	2·944	1·72
	0·38	8·738	1·46		4·8	2·944	1·72
	358·0	0·206	10·65		14·6	2·119	1·00
AO-70	1·49	0·642	0·84	AO-83	11·8	2·159	0·63
	31·5	0·626	2·22		14·6	2·119	1·00
	125·0	0·746	8·58		0·218	0·166	0·38
AO-71	12·5	1·393	0·49	AO-85	0·218	0·166	0·38
	22·4	1·295	1·84		0·218	0·166	0·38

Aqueous–Organic Systems (cont.)

System number	Upper concentration in x-phase	Slope M	% error in M	System number	Upper concentration in x-phase	Slope M	% error in M
	58·6	0·369	3·44		14·8	2·804	3·46
AO-86	51·9	0·352	2·89	AO-93	14·8	2·804	3·46
	51·9	0·352	2·89		14·8	2·804	3·46
	58·7	0·869	1·66		67·5	0·607	0·96
AO-87	58·7	0·869	1·66	AO-94	25·5	0·642	0·27
	58·7	0·869	1·66		67·5	0·607	0·96
	13·6	3·286	3·78		0·02947	5·894	1·01
AO-88	13·6	3·286	3·78	AO-96	0·02947	5·894	1·01
	13·6	3·286	3·78		0·02947	5·894	1·01
	15·0	2·856	3·81		0·0358	25·143	0·94
AO-91	15·0	2·856	3·81	AO-99	0·0255	24·568	0·79
	15·0	2·856	3·81		0·0358	25·143	0·94

Organic–Organic Systems

System number	Upper concentration in x-phase	Slope M	% error in M	System number	Upper concentration in x-phase	Slope M	% error in M
	0·0490	0·759	3·91		9·0	1·683	3·70
OO-1	0·0121	0·663	0·40	OO-8	3·7	1·883	0·93
	0·0121	0·663	0·40		6·4	1·794	1·67
	24·2	1·612	4·00		17·2	1·904	1·33
OO-2	6·2	1·800	1·08	OO-9	11·2	1·983	0·66
	17·5	1·780	1·94		17·2	1·904	1·33
	32·0	1·452	1·34		20·0	1·683	6·72
OO-4	32·0	1·452	1·34	OO-10	5·9	2·243	1·88
	32·0	1·452	1·34		5·9	2·243	1·88
	22·9	1·653	4·14		0·258	2·649	5·00
OO-5	15·5	1·827	2·28	OO-12	0·0726	3·249	1·28
	15·5	1·827	2·28		0·0726	3·249	1·28
	31·85	1·486	2·22		0·3390	2·102	9·87
OO-6	31·85	1·486	2·22	OO-13	0·0235	3·068	0·18
	31·85	1·486	2·22		0·1330	2·948	1·61
	45·0	1·140	4·90		39·0	6·397	7·32
OO-7	9·7	0·499	1·18	OO-14	30·0	1·547	1·96
	9·7	0·499	1·18		30·0	1·547	1·96

Organic–Organic Systems (cont.)

System number	Upper concentration in *x*-phase	Slope M	% error in M	System number	Upper concentration in *x*-phase	Slope M	% error in M
	31·8	1·715	1·30		0·297	2·903	1·98
OO-16	4·98	1·608	0·09	OO-34	0·280	2·794	0·51
	31·8	1·715	1·30		0·297	2·903	1·98
	34·6	1·218	2·52		3·545	1·483	5·92
OO-17	19·9	1·381	0·63	OO-35	2·328	1·836	3·51
	30·7	1·280	1·73		2·328	1·836	3·51
	32·6	1·215	2·54		0·557	0·309	6·41
OO-18	17·1	1·398	0·22	OO-36	0·205	0·2000	1·16
	26·6	1·286	1·50		0·359	0·2113	1·58
	19·6	1·073	3·36		37·6	2·459	7·70
OO-19	7·8	1·239	3·35	OO-37	5·53	3·753	0·52
	19·6	1·073	3·36		9·57	3·656	1·22
	51·1	0·708	4·19		17·9	1·664	4·88
OO-21	13·6	0·488	1·54	OO-38	7·73	2·089	2·55
	13·6	0·488	1·54		7·73	2·089	2·55
	76·3	0·223	7·90		18·62	4·550	2·83
OO-22	37·8	0·169	3·06	OO-39	3·49	5·334	0·63
	37·8	0·169	3·06		12·54	4·777	1·78
	42·6	0·968	0·88		0·146	0·337	11·96
OO-24	24·5	0·987	0·28	OO-41	0·052	0·1938	1·55
	42·6	0·968	0·88		0·052	0·1938	1·55
	78·7	0·0766	2·45		0·069	16·403	17·81
OO-28	57·8	0·0801	1·00	OO-42	0·010	42·848	2·82
	57·8	0·0801	1·00		0·010	42·848	2·82
	0·9063	7·011	5·30		0·027	16·232	4·68
OO-29	0·5061	8·231	2·34	OO-43	0·020	17·515	2·16
	0·5061	8·231	2·34		0·020	17·515	2·16
	1·3300	6·240	5·71		0·098	3·298	8·50
OO-31	0·9569	7·083	3·50	OO-44	0·019	5·277	2·53
	0·9569	7·083	3·50		0·019	5·277	2·53
	14·0	0·440	2·10		0·100	4·842	17·05
OO-32	14·0	0·440	2·10	OO-47	0·013	11·352	3·15
	14·0	0·440	2·10		0·013	11·352	3·15
	0·128	6·406	1·74				
OO-33	0·024	7·206	1·13				
	0·128	6·406	1·74				

Liquid–Gas Systems

System number	Upper concentration in x-phase	Slope M	% error in M	System number	Upper concentration in x-phase	Slope M	% error in M
	1·887	951·2	0·77		0·09506	2415	0·98
LG-1	1·688	935·0	0·53	LG-26	0·09506	2415	0·98
	1·887	951·2	0·77		0·09506	2415	0·98
	2483	0·722	0·77		37·5	7·963	2·22
LG-2	2221	0·710	0·54	LG-27	37·5	7·963	2·22
	2483	0·722	0·77		37·5	7·963	2·22
	1·249	2275	2·85		36·47	18·794	1·08
LG-5	0·544	1443	2·15	LG-28	36·47	18·794	1·08
	0·544	1443	2·15		36·47	18·794	1·08
	1640·7	1·723	2·84		0·3641	2538	4·81
LG-6	714·6	1·098	2·15	LG-29	0·1453	2111	0·00
	714·6	1·098	2·15		0·2061	2187	2·13
	0·816	3867	3·84		0·3248	2641	3·65
LG-11	0·329	2408	1·05	LG-30	0·1356	2341	0·71
	0·329	2408	1·05		0·2402	2455	1·69
	1065·5	2·955	3·84		0·3185	2848	3·74
LG-12	429·6	1·844	1·05	LG-31	0·1246	2487	0·45
	429·6	1·844	1·05		0·1928	2551	1·37
	26·87	1·844	2·98		0·2901	2994	2·86
LG-13	26·87	1·844	2·98	LG-32	0·1243	2648	0·00
	26·87	1·844	2·98		0·1923	2761	1·72
	23·43	2·946	0·40		0·210	4994	2·72
LG-14	13·38	2·996	0·15	LG-34	0·210	4994	2·72
	23·43	2·946	0·40		0·210	4994	2·72
	29·6	2·614	0·23		0·233	4828	3·20
LG-15	13·1	2·675	0·13	LG-35	0·233	4828	3·20
	29·6	2·614	0·23		0·233	4828	3·20
	25·6	4·774	1·02		0·2471	4299	3·30
LG-16	17·2	5·004	0·56	LG-36	0·2471	4299	3·30
	25·6	4·774	1·02		0·2471	4299	3·30
	26·6	4·500	0·98		82·17	2·725	1·05
LG-17	16·6	4·817	0·21	LG-38	3·08	3·263	0·83
	26·6	4·600	0·98		82·17	2·725	1·05
	16·15	9·935	0·35		78·39	6·396	1·01
LG-18	15·91	9·908	0·32	LG-39	29·51	6·865	0·68
	16·15	9·935	0·35		78·39	6·396	1·01
	12·87	10·890	1·25				
LG-19	12·87	10·890	1·25				
	12·87	10·890	1·25				

APPENDIX 6i
PARAMETERS OF LINES THROUGH MEAN VALUES $\bar{x}$ AND $\bar{y}$

Only systems having a minimum error in the slope of less than 4% are shown. Three concentration ranges are quoted for each system, these being from top to bottom;

(i) the full concentration range,
(ii) the range over which the error in the slope is a minimum,
(iii) the widest range over which the error in the slope is less than about 2%.

Aqueous–Organic Systems

System number	Upper concentration in x-phase	Lower concentration in x-phase	$\bar{x}$	$\bar{y}$	Error in $\bar{x}$ and $\bar{y}$	Slope m	% error in m
		1·0	5·50	4·039	0·051	0·869	2·72
AO-1	10·0	8·0	9·0	7·20	0·000	1·010	1·00
		3·0	6·5	4·85	0·031	0·923	1·99
		0·88	4·986	6·591	0·042	1·235	1·57
AO-2	12·0	6·35	8·85	11·397	0·024	1·149	1·34
		0·88	4·986	6·591	0·042	1·235	1·57
		2·12	12·387	13·568	0·218	0·954	4·49
AO-3	22·8	8·35	14·44	15·72	0·058	0·879	1·49
		8·35	14·44	15·72	0·058	0·879	1·49
		0·11	9·902	9·086	0·104	0·865	1·68
AO-4	28·5	12·0	19·405	17·378	0·054	0·800	1·46
		3·25	14·493	13·205	0·027	0·835	1·59

		16·3	27·860	14·340	0·157	0·585	3·78
AO-5	40·1	27·5	33·533	17·767	0·057	0·517	2·42
		27·5	33·533	17·767	0·057	0.517	2.42
		3·6	9·86	8·40	0·124	0·988	4·18
AO-6	15·4	10·2	12·90	11·37	0·042	1·114	2·67
		10·2	12·90	11·37	0·042	1·114	2·67
		2·3	7·64	6·38	0·087	0·935	3·59
AO-7	12·5	2·3	7·64	6·38	0·087	0·935	3·59
		2·3	7·64	6·38	0·087	0·935	3·59
		6·65	15·55	15·46	0·180	1·071	3·12
AO-9	26·6	16·50	21·43	21·90	0·062	0·940	2·18
		16·50	21·43	21·90	0·062	0·940	2·18
		5·45	18·91	18·48	0.049	0·985	0·76
AO-10	31·0	15·15	22·83	21·83	0·029	0·973	0·58
		5·45	18·91	18·48	0·049	0·985	0·76
		0·00306	0·01206	0·01079	0·00011	0·822	2·41
AO-12	0·02539	0·00931	0·01589	0·01406	0·00008	0·782	1·98
		0·00931	0·01589	0·01406	0·00008	0·782	1·98
		0·118	1·061	0·622	0·014	0·667	2·97
AO-13	2·87	0·860	1·729	1·048	0·012	0·718	2·69
		0·3462	1·196	0·703	0·013	0·680	2·78
		0·942	10·786	8·418	0·095	0·859	2·07
AO-14	22·4	13·42	17·428	14·130	0·023	0·985	1·04
		8·73	15·588	12·429	0·044	0·949	1·48
		1·02	16·129	0·965	0·070	0·0788	7·62
AO-17	34·55	25·9	30·300	2·163	0·015	0·1236	3·54
		25·9	30·300	2·163	0·015	0·1236	3·54

Aqueous–Organic Systems (cont.)

System number	Upper concentration in x-phase	Lower concentration in x-phase	$\bar{x}$	$\bar{y}$	Error in $\bar{x}$ and $\bar{y}$	Slope m	% error in m
AO-21	24·5	3·0	13·620	21·647	0·189	1·282	3·197
		6·62	15·744	24·554	0·108	1·220	2·21
		6·62	15·744	24·554	0·108	1·220	2·21
AO-22	10·0	1·0	5·500	7·022	0·045	1·353	1·95
		6·0	8·000	10·382	0·021	1·518	1·78
		1·0	5·500	7·022	0·045	1·353	1·195
AO-24	22·0	0·911	7·970	6·413	0·113	0·931	2·67
		8·44	13·835	11·718	0·040	1·012	1·02
		3·57	9·624	7·794	0·075	0·967	1·85
AO-27	0·00111	0·00039	0·00077	0·05561	0·00001	69·582	3·14
		0·00039	0·00077	0·05561	0·00001	69·582	3·14
		0·00039	0·00077	0·05561	0·00001	69·582	3·14
AO-29	0·00088	0·00018	0·00060	0·00629	0·00001	16·646	4·86
		0·00060	0·00075	0·00888	0	21·784	2·99
		0·00060	0·00075	0·00888	0	21·784	2·99
AO-32	6·5	3·0	4·967	32·833	0·020	7·024	1·65
		4·7	5·700	38·075	0·006	6·588	0·92
		3·0	4·967	32·833	0·020	7·024	1·65
AO-33	11·0	1·9	5·583	23·138	0·238	3·669	7·55
		6·1	8·467	33·933	0·051	2·589	2·70
		6·1	8·467	33·933	0·051	2·589	2·70
AO-37	0·1559	0·0303	0·0868	0·5812	0·00076	7·277	1·89
		0·0303	0·0868	0·5812	0·00076	7·277	1·89
		0·0303	0·0868	0·5812	0·00076	7·277	1·89

		0·00553	0·0084	0·6640	0·00002	85·467	1·05
AO-38	0·01125	0·00553	0·0024	0·6640	0·00002	85·467	1·05
		0·00553	0·0024	0·6640	0·00002	85·467	1·05
		0·00454	0·05350	0·03426	0·00016	0·688	0·57
AO-39	0·1824	0·01068	0·06036	0·03880	0·00015	0·692	0·54
		0·00454	0·05350	0·03426	0·00016	0·688	0·57
		0·00099	0·0717	0·0383	0·0002	0·564	0·36
AO-40	0·5899	0·00629	0·0754	0·0403	0·0002	0·565	0·35
		0·00099	0·0717	0·0383	0·0002	0·564	0·36
		3·8	11·917	5·550	0·420	0·830	12·03
AO-41	19·8	13·9	16·733	9·467	0·015	1·288	0·77
		13·9	16·733	9·467	0·015	1·288	0·77
		8·0	19·557	16·957	0·306	1·357	5·50
AO-43	28·2	13·6	21·483	19·233	0·143	1·489	3·14
		13·6	21·483	19·233	0·143	1·489	3·14
		0·170	0·9999	0·4393	0·0026	0·453	0·81
AO-47	2·678	0·170	0·9999	0·4393	0·0026	0·453	0·81
		0·170	0·9999	0·4393	0·0026	0·453	0·81
		0·00147	0·00742	0·03278	0·00034	6·525	7·74
AO-50	0·01538	0·00989	0·01247	0·06420	0·00002	8·957	1·09
		0·00989	0·01247	0·06520	0·00002	8·957	1·09
		0·0016	0·1407	2·1096	0·0047	14·977	3·66
AO-51	0·379	0·2730	0·3200	4·8190	0·0007	20·548	1·47
		0·2730	0·3200	4·8190	0·0007	20·548	1·47
		0·008	0·176	2·777	0·006	18·676	3·67
AO-53	0·452	0·332	0·387	6·667	0·001	16·595	2·12
		0·332	0·387	6·667	0·001	16·595	2·12

Aqueous–Organic Systems (cont.)

System number	Upper concentration in x-phase	Lower concentration in x-phase	$\bar{x}$	$\bar{y}$	Error in $\bar{x}$ and $\bar{y}$	Slope m	% error in m
AO-54	12·5	0·52	7·484	6·903	0·036	0·951	1·37
		9·60	10·800	10·123	0·0004	1·045	0·05
		0·52	7·484	6·903	0·036	0·951	1·37
AO-55	22·38	2·18	10·089	5·097	0·123	0·572	3·69
		2·18	10·089	5·097	0·123	0·572	3·69
		2·18	10·089	5·097	0·123	0·572	3·69
AO-56	94·60	2·19	27·535	4·078	0·040	0·1487	0·86
		22·30	50·440	7·537	0·018	0·1463	0·38
		2·19	27·535	4·078	0·040	0·1487	0·86
AO-65	0·212	0·045	0·128	0·331	0·002	3·137	4·33
		0·150	0·181	0·502	0·0006	2·291	2·68
		0·045	0·045	0·128	0·331	3·137	4·33
AO-67	1·87	0·45	1·126	1·586	0·002	10·446	0·50
		0·61	1·238	12·748	0·002	10·499	0·48
		0·45	1·126	11·586	0·002	10·446	0·50
AO-68	1·97	0·44	1·027	9·820	0·009	9·508	1·72
		0·44	1·027	9·820	0·009	9·508	1·72
		0·44	1·029	9·820	0·009	9·508	1·72
AO-69	0·38	0·14	0·294	2·561	0·004	9·257	6·02
		0·32	0·343	3·038	0·0006	10·436	2·61
		0·32	0·343	3·038	0·0006	10·436	2·61
AO-70	358·0	0·62	76·045	21·315	1·987	0·1630	12·29
		163·0	244·667	47·333	0·176	0·0876	2·44
		163·0	244·667	47·333	0·176	0·0876	2·44

AO-73	81·6	2·14	32·213	11·276	0·086	0·381	0·96
		2·14	32·213	11·276	0·086	0·381	0·96
		2·14	32·213	11·276	0·086	0·381	0·96
AO-78	61·0	7·6	35·133	17·900	0·230	0·628	2·54
		7·6	35·133	17·900	0·230	0·628	2·54
		7·6	35·133	17·900	0·230	0·628	2·54
AO-83	14·6	6·2	10·750	22·850	0·095	2·025	3·74
		11·8	13·200	13·200	0·013	1·536	1·34
		11·8	13·200	27·867	0·013	1·536	1·34
AO-85	0·218	0·0176	0·1077	0·0179	0·0001	0·1662	0·75
		0·0176	0·1077	0·0179	0·0001	0·1662	0·75
		0·0176	0·1077	0·0179	0·0001	0·1662	0·75
AO-86	58·6	8·4	34·063	12·088	0·255	0·427	3·86
		8·4	34·063	12·088	0·255	0·427	3·86
		8·4	34·063	12·088	0·255	0·427	3·86
AO-87	58·7	6·8	30·556	26·433	0·402	0·882	3·52
		22·1	39·800	35·050	0·290	0·795	3·46
		22·1	39·800	35·050	0·290	0·795	3·46
AO-88	13·6	5·9	10·350	33·133	0·134	4·313	5·29
		11·4	12·567	42·467	0·010	2·815	1·17
		11·4	12·567	42·467	0·010	2·185	1·17
AO-89	11·6	5·6	8·675	26·950	0·084	5·340	3·76
		5·6	8·675	26·950	0·084	5·340	3·76
		5·6	8·675	26·950	0·084	5·340	3·76
AO-90	13·0	7·2	10·500	26·650	0·042	4·512	1·96
		10·0	11·600	31·533	0·012	4·728	0·96
		7·2	10·500	26·650	0·042	4·512	1·96

Aqueous–Organic Systems (cont.)

System number	Upper concentration in x-phase	Lower concentration in x-phase	$\bar{x}$	$\bar{y}$	Error in $\bar{x}$ and $\bar{y}$	Slope m	% error in m
AO-93	14·8	6·6	11·400	31·267	0·164	3·603	6·08
		12·6	13·800	39·600	0·011	2·178	1·28
		12·6	13·800	39·600	0·011	2·178	1·28
AO-94	67·5	2·81	27·369	16·883	0·136	0·590	1·28
		2·81	27·369	16·883	0·136	0·590	1·28
		2·81	27·369	16·883	0·136	0·590	1·28
AO-96	0·02947	0·00135	0·01447	0·08288	0·00005	6·305	0·53
		0·00519	0·01646	0·09515	0·00003	6·376	0·41
		0·00135	0·01447	0·08288	0·00005	6·305	0·53
AO-98	0·46	0·149	0·302	0·126	0·007	1·199	9·63
		0·300	0·369	0·202	0·001	1·531	2·37
		0·300	0·369	0·202	0·001	1·531	2·37
AO-99	0·0358	0·0198	0·0279	0·6988	0·0001	26·945	1·84
		0·0198	0·0279	0·6988	0·0001	26·945	1·84
		0·0198	0·0279	0·6988	0·0001	26·945	1·84

Organic–Organic Systems

System number	Upper concentration in x-phase	Lower concentration in x-phase	$\bar{x}$	$\bar{y}$	Error in $\bar{x}$ and $\bar{y}$	Slope m	% error in m
OO-1	0·049	0·009	0·0271	0·0198	0·0005	0·884	5·49
		0·020	0·0327	0·0240	0·0002	0·998	2·23
		0·020	0·0327	0·0240	0·0002	0·998	2·23
OO-4	32·0	8·0	20·200	29·425	0·285	1·430	3·93
		8·0	20·200	20·425	0·285	1·430	3·93
		8·0	20·200	29·425	0·285	1·430	3·93
OO-16	31·8	4·35	15·803	27·283	0·207	1·680	2·66
		4·35	15·803	27·283	0·207	1·680	2·66
		4·35	15·803	27·283	0·207	1·680	2·66
OO-19	19·6	6·2	11·740	12·920	0·042	0·899	1·27
		6·2	11·740	12·920	0·042	0·899	1·27
		6·2	11·740	12·920	0·042	0·899	1·27
OO-24	42·6	10·0	25·700	24·967	0·193	0·953	2·10
		10·0	25·700	24·967	0·193	0·953	2·10
		10·0	25·700	24·967	0·193	0·953	2·10
OO-32	14·0	1·2	5·503	2·453	0·065	0·432	3·60
		3·775	7·942	3·579	0·006	0·408	0·34
		3·775	7·942	3·579	0·006	0·408	0·34
OO-33	0·128	0·010	0·0521	0·3446	0·0008	5·969	2·24
		0·045	0·0842	0·5411	0·0006	5·385	2·06
		0·045	0·0842	0·5411	0·0006	5·385	2·06
OO-39	18·62	1·64	8·884	41·543	0·211	4·062	4·49
		12·00	14·387	63·003	0·004	3·212	0·15
		9·25	12·380	56·020	0·075	3·358	2·33

Liquid–Gas Systems

System number	Upper concentration in x-phase	Lower concentration in x-phase	$\bar{x}$	$\bar{y}$	Error in $\bar{x}$ and $\bar{y}$	Slope m	% Error in m
LG-1	1·887	0·997	1·476	1397	0·005	1058	1·60
		1·778	1·837	1788	0·001	1508	0·00
		0·997	1·476	1397	0·005	1058	1·60
LG-2	2483	1312	1942	1397	3·87	0·803	1·60
		1312	1942	1397	3·87	0·803	1·60
		1312	1942	1397	3·87	0·803	1·60
LG-3	1·0855	0·414	0·759	900	0·003	1806	1·41
		0·871	0·978	1300	0·002	1864	0·00
		0·414	0·759	900	0·003	1806	1·41
LG-4	1428	544·6	998	900	3·32	1·372	1·41
		1005	1216	1200	0·02	1·417	0·02
		544·6	998	900	3·32	1·372	1·41
LG-5	1·249	0·497	0·927	1999	0·006	3585	2·57
		0·935	1·095	2598	0·000	4436	0·00
		0·707	1·005	2249	0·003	3994	1·88
LG-6	1640·7	652·9	1218	1999	6·9	2·724	2·57
		1292·6	1468	2698	0·29	3·407	0·28
		928·8	1320	2249	3·8	3·039	1·91
LG-7	0·8149	0·2945	0·576	900	0·007	2351	4·00
		0·6694	0·742	1300	0·000	2749	0·00
		0·5083	0·665	1100	0·001	2617	1·68

		386·9	756·7	900	7·8	1·787	3·99
LG-8	1070·4	879·3	974·9	1300	0·02	2·093	0·03
		667·7	873·8	1100	2·1	1·992	1·69
		0·2121	0·4415	900	0·006	2906	4·45
LG-9	0·6313	0·3879	0·5141	1100	0·002	3287	2·87
		0·3879	0·5141	1100	0·002	3287	2·87
		277·9	578·5	900	7·3	2·214	4·46
LG-10	827·1	679·5	754·3	1300	1·4	2·710	2·43
		679·5	754·3	1300	1·4	2·710	2·43
		0·315	0·5806	2112	0·004	5849	2·76
LG-11	0·816	0·706	0·7592	3208	0·000	7645	0·00
		0·433	0·6354	2403	0·002	6377	1·56
		411·3	758·1	2112	5·7	4·477	2·86
LG-12	1065·5	921·7	991·3	3208	0·1	5·850	0·30
		840·9	954·3	3006	1·3	5·589	1·75
		16·33	21·894	39·714	0·010	2·647	0·30
LG-13	26·87	16·33	21·894	39·714	0·010	2·647	0·30
		16·33	21·894	39·714	0·010	2·647	0·30
		9·98	16·851	49·778	0·031	2·834	0·75
LG-14	23·43	20·39	22·033	64·333	0·004	2·633	0·37
		9·98	16·851	49·778	0·031	2·834	0·75
		11·2	20·891	54·727	0·010	2·551	0·19
LG-15	29·6	11·2	20·891	54·727	0·010	2·551	0·19
		11·2	20·891	54·727	0·010	2·551	0·19
		8·5	16·565	79·765	0·130	4·368	2·41
LG-16	25·6	22·5	24·325	111·500	0·005	3·557	0·40
		10·9	18·136	87·214	0·093	4·096	1·98

Liquid–Gas Systems (cont.)

System number	Upper concentration in x-phase	Lower concentration in x-phase	$\bar{x}$	$\bar{y}$	Error in $\bar{x}$ and $\bar{y}$	Slope m	% error in m
LG-17	26·6	8·5	17·188	79·765	0·121	4·203	2·18
		24·7	25·867	113·667	0·007	3·147	0·88
		15·5	21·210	97·100	0·063	3·653	1·74
LG-18	16·15	6·35	11·127	110·130	0·027	10·377	0·84
		6·84	11·576	115·000	0·023	10·241	0·77
		6·35	11·127	110·130	0·027	10·377	0·84
LG-26	0·09506	0·01122	0·05357	127·09	0·00040	2585	1·56
		0·01122	0·05357	127·09	0·00040	2585	1·56
		0·01122	0·05357	127·09	0·00040	2585	1·56
LG-27	37·5	0·534	17·978	137·94	0·392	8·410	2·85
		31·52	35·097	287·67	0·002	10·533	0·07
		21·03	30·560	242·60	0·102	9·946	1·67
LG-28	36·47	0·525	15·589	286·49	0·127	19·414	1·02
		21·72	29·098	550·00	0·002	20·338	0·04
		0·525	13·589	286·49	0·127	19·414	1·02
LG-29	0·3641	0·1103	0·2180	528·72	0·0049	3065	5·43
		0·2061	0·2781	701·40	0·0025	3433	3·87
		0·2061	0·2781	701·40	0·0025	3433	3·87
LG-31	0·3185	0·0920	0·1942	535·72	0·004	3281	5·39
		0·1928	0·2515	713·43	0·0016	3767	3·09
		0·1928	0·2515	713·43	0·0016	3767	3·09

LG-32	0·2901	0·0947	0·1840	536·60	0·0024	3452	3·51
		0·947	0·1840	536·60	0·0024	3452	3·51
		0·0947	0·1840	536·60	0·0024	3452	3·51
LG-33	0·1893	0·0179	0·0978	670·03	0·0005	4456	0·81
		0·0378	0·1111	729·75	0·0000	4442	0·00
		0·0179	0·0978	670·03	0·0005	4456	0·81
LG-34	0·210	0·023	0·1219	626·91	0·0000	4327	0·00
		0·023	0·1219	626·91	0·0000	4327	0·00
		0·023	0·1219	626·91	0·0000	4327	0·00
LG-35	0·233	0·0300	0·1274	637·61	0·0000	4106	0·00
		0·1633	0·1985	929·33	0·0000	4168	0·00
		0·0300	0·1274	637·61	0·0000	4168	0·00
LG-36	0·2471	0·0621	0·1448	604·33	0·0042	4794	6·17
		0·1855	0·2180	941·77	0·0000	4496	0·00
		0·0720	0·1585	684·98	0·0005	4332	0·88
LG-37	0·1950	0·0175	0·1080	684·24	0·0007	4248	1·14
		0·1365	0·1647	925·17	0·0000	3885	0·00
		0·0175	0·1080	684·24	0·0007	4248	1·14
LG-38	82·17	0·73	30·497	84·611	0·352	2·674	1·24
		0·73	30·497	84·611	0·352	2·674	1·24
		0·73	30·497	84·611	0·352	2·674	1·24
LG-39	78·39	0·69	24·265	158·53	0·288	6·271	1·13
		45·90	62·397	396·27	0·097	5·812	0·74
		0·69	24·265	158·53	0·288	6·271	1·13

APPENDIX 6j
PHYSICAL PROPERTIES OF SOLVENTS AND SOLUTES

KEY TO TOXICITY RATINGS

- U Unknown toxicity
- T Slight toxicity
- MT Medium toxicity, producing irreversible as well as reversible changes in the human body, but not of such severity as to threaten life or produce serious permanent physical impairment *except* in special cases of exposure
- VT Severe toxicity, producing irreversible changes of such severity as to threaten life
- R Radioactive

CONDITIONS OF MEASUREMENT

Specific Gravity. The figures following the S. G. indicate the temperature of measurement and the temperature of the water to which it is referred.

Viscosities are given at 20 °C (except where indicated).

Boiling points are given at 760 mm Hg pressure (except where indicated).

Solubility refers to the weight of substance in grams which 100 g of water will dissolve at 20 °C (except where indicated). For gases the volume in cc at 760 mm Hg pressure is sometimes given instead.

Substance	S.G.	Viscosity cp at 20°C	F.pt. °C	B.pt. °C at 760 mm Hg pressure	Flash pt. °F		Explosive range vol. % in air	Tox-icity	Solubility in water
					Open cup	Closed cup			
Acetaldehyde	0·7827 20/20	1·3	−123·5	20·8	−58	−36	4·0−57.0	MT	∞
Acetic acid	1·0507 20/20	1·25	16·7	118·1	135	104	4·0−	T	∞
Acetone	0·7972 15/4	0·33	−94·6	56·5	15	0	2·15−13·0	T	∞
Ammonia	0·000771 0/4	0·01	−77·7	33·4	–	–	16·0–25·0	MT	89·9 (0°C)
n-Amyl acetate	0·879 20/20	0·9	−70·4	148·4 (737 mm Hg)	–	77	1·10–	T	very slight
n-Amyl alcohol	0·8168 20/20	4·4	−78·5	137·8	120–136	91	1·2–	MT	2·7 (22°C)
Aniline	1·02 20/4	4·3	−6·2	184·4	–	168	–	VT	3·6 (18°C)
Benzene	0·8794 20/4	0·65	5·5	80·1	–	12	1·4–	MT	0·07 (22°C)
Benzoic acid	1·266 15/4	–	121·7	249	–	250	–	T	0·2 (17°C)
1-Butanol	0·8098 20/4	2·9	−79·9	117·5	114	–	1·7–18·0	T	9 (15°C)
n-Butyl acetate	0·882 20/20	0·68	−76·3	126	92	–	1·7–15·0	T	0·7
1-Butyraldehyde	0·817 20/4		−99	75·7	–	20	–	U	4
n-Butyric acid	0·9590 20/20	1·6	−4·7	163·5	170	–	–	T	∞
n-Caproic acid	0·9295 20/20		−1·5	205·0	215	–	–	T	1·1
Carbon dioxide	–	0·014	−56·6	−78·2	–	–	–	T	90·1 cc
Carbon tetrachloride	1·597 20/4	1·0	−22·6	76·8	–	–	–	MT	0·08
Chlorobenzene	1·113 15/15	0·9	−45·2	131·7	–	85	1·8–9.6	T	0·049
Chloroform	1·4985 15/4	0·57	−63·5	61·3	–	–	–	MT	0·82
Cyclohexane	0·7791 20/4	1·0	6·5	80·7	–	1	1·3–8·35	T	insoluble
Cyclohexanol	0·9449 25/4	57 (25°C)	23·9	161·5	–	154	–	T	3·6
Cyclohexene	0·8102 20/4		−103·7	83·3	–	21	–	T	very slight
n-Decane	0·730 20/4		−29·7	174·1	–	115	0·67–2·6	U	insoluble
Dibutyl carbitol	0·8853			254·6	260	–	–	U	
Diethylamine	0·7108 20/20		40	55·5	0	–	–	MT	very soluble

(cont.)

Substance	S.G.	Viscosity cp at 20 °C	F.pt. °C	B.pt. °C at 760 mm Hg pressure	Flash pt. °F		Explosive range vol. % in air	Tox-icity	Solubility in water
					Open cup	Closed cup			
Diethylene glycol	1·1184 20/20	−38·9		245·8	275	—	—	T	soluble
Dimethyl ether	0·661		−138·5	−23·7	—	−42	3·45–26·7	U	3700 cc (18 °C)
Ethanol	0·7893 20/4	1·2	−112	78·3	61	—	3·28–19·0	T	∞
Ethyl acetate	0·8946 25/4	0·48	−82·4	77·2	40	24	2·2–11·5	T	8·5 (15 °C)
Ethyl benzene	0·8669 20/4	0·74	−94·4	136·2	85	—	—	MT	0·01 (15 °C)
Ethyl butyrate	0·8788		−93·3	121·0	—	78	—	U	0·68 (25 °C)
Ethyl ether	0·7135 20/4	0·24	−116·3	34·6	—	40	1·85–36·5	MT	7·5
Ethyl propionate	0·895 15·5/4	0·54	−72·6	99	—	54	—	U	2·4
Ethylene glycol	1·113 25/25	23	−15·6	197·5	—	232	3·2–	T	∞
Formic acid	1·2267 15/4	1·9	8·6	100·8	156	—	—	MT	∞
Furfural	1·161 20/20		−38·7	161·7 (764 mm Hg)	—	140	2·1 at 125 °C	MT	9·1 (13 °C)
Glycerol	1·260 20/4	1500	17·9	290	349	—	—	T	∞
Heptadecanol	0·8469 20/20		54	309	310	—	—	U	
n-Heptane	0·684 20/4	0·42	−90·6	98·5	—	25	1·2–6·7	T	0·005 (15 °C)
Hexamethyl-disiloxane	0·7606 25/25			99·5	30	—	—	T	
n-Hexane	0·6603 20/4	0·33	−94	68·7	−10	—	1·2–6·9	T	0·014 (15 °C)
1-Hexanol	0·8186 20/4		−51·6	157·2	165	—	—	U	0·6
Hydrogen bromide	0·0035 0/4	0·017	−86	−66·5	—	—	—	VT	221 (0 °C)
Hydrogen chloride	0·00164 0/4	0·014	−111	−84·8	—	—	—	VT	82·3 (0 °C)
Hydrogen iodide	0·00566 0/4	0·018	−50·8	−35·4 (4 atm)	—	—	—	VT	very soluble

Isoamyl acetate	0·876			142·0	77	–	1·0 at 212 °F	MT	0·3 (15 °C)
Isoamyl alcohol	0·813 15/4		−117·2	132	–	109	1·2–	MT	2 (14 °C)
Isophorone	0·9229			215·2	205	–	–	VT	very slight
Isovaleric acid	0·931 20/20		−37·6	176	–	–	–	–	4·2
Kerosene	0·8–1·0	2·4		175–325	–	100–165	1·16–6·0	T	insoluble
Lactic acid	1·249 15/4		16·8	122 (15 mm Hg)	–	–	–	U	∞
Malonic acid	1·631 15/4		130 (decom-poses)	–	–	–	–	–	138 (16 °C)
Methane	0·0000717 0/4		−182·6	−161·5	–	–	5·3–14·0	T	0·4 cc
Methanol	0·7913 20/4	0·6	−97	64·8	65	–	6·0–36·5	MT	∞
Methyl acetate	0·9244	0·41	−98·7	57·8	14	–	4·1–13·9	MT	33 (22 °C)
Methyl 1-butyl ketone	0·830 0/4			127·2	95	–	1·22–8·0	MT	
Methyl butyrate	0·898	0·6	−95	102·3	–	57	–	T	1·7
Methyl cyclohexane	0·769 20/4		−126·3	100·3	–	25	1·15–	T	insoluble
Methyl cyclo-hexanone	0·925 15/5			160–170	145	118	–	MT	
Methyl ethyl ketone	0·8062 20/20	0·43	−85·9	79·6	22	–	1·81–11·5	T	35 (10 °C)
Methyl isobutyl carbinol	0·813 20/4		−117·2	131·6	–	106	–	T	
Methyl isobutyl ketone	0·8024 20/20			115·1	75	–	–	MT	
Nicotine	1·009 20/4		−80	247·3	–	–	0·75–4·0	VT	soluble
Nitrobenzene	1·1987 25/4	2·1	5·7	210·9	–	190	1·8 at 200 °F	VT	0·19

(cont.)

Substance	S.G.	Viscosity cp at 20°C	F.pt. °C	B.pt. °C at 760 mm Hg pressure	Flash pt. °F		Explosive range vol. % in air	Tox-icity	Solubility in water
					Open cup	Closed cup			
Nitromethane	1·130 20/4			101	–	95	7·3–	T	
1-Nitropropane	1·003 20/20			132	120	–	–	MT	
Octyl acetate	0·872 20/4		–38·5	200	180	–	–	T	insoluble
Olive oil	0·910	84		–	–	437	–	–	insoluble
Oxalic acid	1·653		101·5	sublimes at 150°C	–	–	– –	VT	soluble
Phenol	1·072		42	181·9	–	175	2·3–7·3	VT	8·2 (15°C)
Propane	0·5853		–187·1	–42·1	–156	–		T	6·5 cc (18°C)
n-Propanol	0·8044 20/4	2·4	–127	97·2	–	59	2·5–13·5	T	∞
Propionic acid	0·992	1·15	–22	141	–	–	–	U	∞
n-Propyl acetate	0·887	0·60	–92·5	101·6	58	–	1·77–8·0	T	1·6 (15°C)
Propylene	0·581 liq. 0/4		–185	–47·7	–	–	–	T	44·6 cc
Propylene glycol	1·0362 25/25			188·2	210	–	2·62–12·55	T	∞
Styrene	0·9074 20/4	–	–31	145·2	–	88	1·1–6·1	MT	very slight
Succinic acid	1·564 15/4	–	190	235 decomposes	–	–	–	T	6·8
Sulphur dioxide	–	0·012	–75·5	–10·0	–	–	–	MT	22·8 (0°C)
Sulphuric acid	1·834	26 (98%)	10·5	330	–	–	–	VT	∞
Tetrachloroethane	1·600 20/4	1·7	–36	146·3	–	–	–	VT	0·29
Toluene	0·866 20/4	0·59	–95	110·4	–	40	1·27–7·0	MT	0·05 (16°C)
Triethylamine	0·7229 25/4	–	–114·8	89·5	20	–	–	U	∞
Uranyl nitrate	2·807	–	60·2	118	–	–	–	VT,R	170·3 (0°C)
n-Valeric acid	0·939 20/4	–	–34·5	187	–	–	–	–	3·3 (16°C)
Vinyl acetate	0·9335 20/4	0·45	less than –60	72·5	–20	18	–	T	2
o-Xylene	0·880 20/4	0·84	–25	144·4	115	63	1·1–7·0	T	insoluble

CHAPTER 7

Conclusions

THIS chapter brings together the results of the previous chapters and shows how they are inter-related. In particular it shows how ordinary counter-current extraction and forward and back extraction using counter-current or cross-current flow may be related by a single form of equation. This major conclusion is stated immediately below and amplified in the subsequent conclusions which are in chronological order.

7.1. Generalised Form of Equation

1. Differential Formula

For a differential contactor, the expression

$$N_T(1 - J) = \ln Q$$

may be used for counter-current mass and heat transfer, and for the determination of optimum conditions in forward and back extraction.

2. Stagewise Formula

For a stagewise contactor, the expression

$$N_S \ln 1/J = \ln Q$$

may be used for counter-current extraction, for the determination of optimum conditions in forward and back extraction if J is replaced by $\sqrt{(J_1 J_2)}$, for cross-current extraction with recirculation if J is replaced by β, and for ordinary cross-current extraction if J is replaced by λ and Q is defined differently.

7.2. Introduction

1. *Absorption and Stripping*

When using the well-established formulae

$$N_S \ln 1/J = N_T (1 - J) = \ln [(1 - J) Q_{2/4} + J]$$

where

$$J = mG/L, \quad Q_{2/4} = \frac{y_{\text{in}} - y(x_{\text{in}})}{y_{\text{out}} - y(x_{\text{in}})} \quad \text{and} \quad N_T = N_G$$

or

$$J^* = L/(mG), \quad Q^*_{2/4} = \frac{x_{\text{in}} - x(y_{\text{in}})}{x_{\text{out}} - x(y_{\text{in}})} \quad \text{and} \quad N^*_T = N_L$$

it is better to use the values corresponding to J or J^* less than unity, rather than to consider whether absorption or stripping is taking place as suggested by previous authors.

2. *Concentration Profiles*

(a) STAGEWISE BASED ON INLET CONCENTRATIONS

When the inlet concentrations are known stagewise concentration profiles may be expressed in terms of the dimensionless concentrations

$$\phi_n = \frac{J(y_{\text{in}} - y_n)}{y_{\text{in}} - y(x_{\text{in}})} = \frac{J^n - 1}{J^N - 1/J}$$

and

$$\xi_{n+1} = \frac{y(x_{n+1}) - y(x_{\text{in}})}{y_{in} - y(x_{in})} = \frac{J^N - J^n}{J^N - 1/J}$$

(b) DIFFERENTIAL BASED ON INLET CONCENTRATIONS

The corresponding differential profiles are

$$\phi = \frac{J(y_{\text{in}} - y)}{y_{\text{in}} - y(x_{\text{in}})} = \frac{e^{-N_G(1-J)h/H} - 1}{e^{-N_L(1-J)} - 1/J}$$

and

$$\xi = \frac{y(x) - y(x_{\text{in}})}{y_{\text{in}} - y(x_{\text{in}})} = \frac{e^{-N_L(1-1/J)} - e^{-N_L(1-1/J)h/H}}{e^{N_L(1-1/J)} - 1/J}$$

(c) STAGEWISE BASED ON INLET AND OUTLET CONCENTRATIONS

When the inlet and outlet concentrations are known stagewise concentration profiles are best expressed in terms of the dimensionless concentrations.

$$\psi_n = \frac{y_{in} - y_n}{y_{in} - y_{out}} = \frac{J^n - 1}{J^N - 1}$$

and

$$\chi_{n+1} = \frac{y(x_{n+1}) - y(x_{in})}{y(x_{out}) - y(x_{in})} = \frac{J^N - J^n}{J^N - 1}$$

so that

$$\psi_n + \chi_{n+1} = 1$$

(d) DIFFERENTIAL BASED ON INLET AND OUTLET CONCENTRATIONS

The corresponding differential profiles are

$$\psi = \frac{y_{in} - y}{y_{in} - y_{out}} = \frac{e^{-N_G(1-J)h/H} - 1}{e^{-N_G(1-J)} - 1}$$

and

$$\chi = \frac{y(x) - y(x_{in})}{y(x_{out}) - y(x_{in})} = \frac{e^{N_L(1-1/J)} - e^{N_L(1-1/J)h/H}}{e^{N_L(1-1/J)} - 1}$$

so that

$$\psi + \chi = 1$$

3. *Cross-current Extraction*

(a) FORMULA RELATING N_C, J AND Q_C

The number of stages in cross-current extraction is given by

$$N_C \ln 1/\lambda = \ln Q_C$$

where

$$\lambda = \frac{J}{1 + J} \quad \text{and} \quad Q_C = \frac{y_{in} - y(x_{in})}{y_{out} - y(x_{in})}$$

(b) COMPARISON WITH COUNTER-CURRENT EXTRACTION

For a given value of J cross-current extraction is always more efficient than counter-current extraction (so that $N_C < N_S$ for a given separation Q, and $Q < Q_C$ for a given number of stages $N_S = N_C$).

7.3. Separation Factors and Basic Formulae

1. Separation Factors

There are fourteen different separation factors as listed in Table 2.1. If negative and reciprocal variations are included there are fifty-six.

2. Inversion

Each of these separation factors may be inverted to give another separation factor already included in the fourteen.

3. Basic Formulae

There are eight basic formulae relating the number of stages and transfer units to the separation factor Q and extraction factor J as listed in Table 2.3. In these formulae a separation factor may be replaced by its inverse if the extraction factor is replaced by its inverse $1/J$.

4. Particular Formula

The separation factor $Q_{2/5}$ defined in Table 2.3 is unchanged on inversion and the number of stages or transfer units are given by

$$N_S \ln 1/J = N_T (1 - J) = \ln \frac{J - Q_{2/5}}{1 - JQ_{2/5}}$$

where

$$J = mG/L \text{ and } N_T = N_G \text{ or } J = L/mG \text{ and } N_T = N_L.$$

5. Recommended Formula

The separation factor

$$Q_{3/4} = \frac{y_{\text{in}} - y(x_{\text{out}})}{y_{\text{out}} - y(x_{\text{in}})}$$

becomes its reciprocal $Q_{4/3}$ on inversion and the number of stages or transfer units are given by

$$N_S \ln 1/J = N_T (1 - J) = \ln Q$$

where

$$Q = Q_{3/4}, \quad J = mG/L \quad \text{and} \quad N_T = N_G$$

or

$$Q = Q_{4/3}, \quad J = L/mG \quad \text{and} \quad N_T = N_L$$

This is the formula recommended for calculating numbers of stages and transfer units.

6. Straight-line Plots

Straight-line relationships between N, J and $Q_{3/4}$ may be easily obtained on log-log or semi-log graph paper with a fairly even gradation of the constant parameter.

7. Graphical Representation

Most of the other formulae relating N, J and Q are difficult to represent graphically because of the uneven gradation of the parameter and the asymptotic nature of the curves.

8. Inaccuracies

Many formulae for calculating numbers of stages and transfer units quoted in the literature are inaccurate. This includes Editions III and IV of the *Chemical Engineers' Handbook.*

9. Heat Transfer

All the formulae listed in Table 2.3 may be used in heat transfer calculations if the analogous quantities listed in Table 2.4 are used.

7.4. Errors in N, J and Q

1. Effect of Errors in J

Absolute and fractional errors in J can give rise to large errors in N and $Q_{3/4}$.

2. Effect of Errors in Q

Absolute and fractional errors in $Q_{3/4}$ usually give smaller errors in N and J.

3. Magnitude of Errors in J and Q

Errors in J are usually small whereas errors in $Q_{3/4}$ may be large, so both must be considered.

4. Effect of Errors in N

Variation in N or the stage efficiency can give rise to large errors in $Q_{3/4}$.

7.5. Forward and Back Extraction Using Counter-current Flow

1. Optimisation

In the optimisation of forward and back extraction using counter-current flow with fixed overall operating conditions $J_1 J_2$, minimising the total number of stages or overall transfer units based on the initial and final solvents for a given separation, leads to the same result as maximising the separation for this total number of stages or transfer units.

2. Differential Extractor

For the differential case the minimum number of transfer units occurs when $N_{L1}/J_1 = N_{G2} = N_{TO}$ and $J_1 = J_2 = J_O = \sqrt{(J_1 J_2)}$ and these are related to the separation $Q_{3/4}$ by

$$N_{TO}(1 - J_O) = \ln Q_{3/4}$$

3. Alternative Formulae

This formula is identical with that given in Conclusion 7.3(*5*). It may be replaced by any of the other formulae for counter-current extraction in a single contactor listed in Table 2.3 if J is replaced by $J_1 J_2$

in that part of the expression containing Q. For example,

$$N_{TO}(1 - J_O) = \ln[(1 - J_1J_2)Q_{2/4} + J_1J_2]$$

when $J_1J_2 = 1$ this becomes

$$N_{TO} = 2(Q_{2/4} - 1)$$

4. *Stagewise Extractor*

For the stagewise case the minimum number of stages occurs for practical purposes when $N_{S1} = N_{S2} = N_{SO}$ and the relationships between N_{SO}, $Q_{3/4}$, J_1J_2 and J_1 or J_2 are given by equations 4(2), (6), (10), (15) and (17). The solution is presented graphically in Figs. 4.4 and 4.5.

When $J_1J_2 = 1$ the governing expressions reduce to $J_1 = J_2 = 1$ and $N_{SO} = 2(Q_{2/4} - 1)$.

5. *Approximate Solution*

When N_{SO} is large (greater than five for most practical purposes) $J_1 = J_2 = \sqrt{(J_1J_2)}$ and N_{SO} is related to the separation $Q_{3/4}$ and overall operating conditions J_1J_2 by

$$N_{SO} \ln 1/\sqrt{(J_1J_2)} = \ln Q_{3/4}$$

The expression becomes more exact as J_1J_2 approaches unity.

6. *Alternative Formulae*

This formula is identical with that given in Conclusion 7.3(5) if $\ln 1/J$ is replaced by $\ln 1/\sqrt{(J_1J_2)}$. Any of the other formulae in Table 2.3 may be used if J is also replaced by J_1J_2 in that part of the expression containing Q. For example,

$$N_{SO} \ln 1/\sqrt{(J_1J_2)} = \ln[(1 - J_1J_2)Q_{2/4} + J_1J_2]$$

when $J_1J_2 = 1$, this becomes

$$N_{SO} = 2(Q_{2/4} - 1)$$

7. *Comparison with Ordinary Counter-current Extraction*

(a) CONSTANT $J = J_1 = J_2$

Under the same conditions of individual extraction $J = J_1 = J_2$, ordinary counter-current extraction and counter-current forward and back extraction give the same separation $Q = Q_O$ from a given number of transfer units $N_T = N_{TO}$, or from a given number of stages $N_S = N_{SO}$ when N_{SO} is large.

(b) CONSTANT $J = J_1 J_2$

Under the same conditions of overall extraction $J = J_1 J_2$, counter-current forward and back extraction always requires more stages or transfer units for a given separation (or gives a poorer separation from a given number of stages or transfer units) than ordinary counter-current extraction.

8. *Operation of Existing Differential Extractor*

For an existing differential contactor in which N_{L1}/J_1 is not equal to N_{G2} the maximum value $Q_{3/4}$ and optimum values of J_1 and J_2 at fixed $J_1 J_2$ are given by

$$\left(\frac{J_1 J_2}{N_1} + \frac{1}{N_2}\right) \ln Q_O = 1 - J_1 J_2$$

and

$$\left(1 - \frac{\ln Q_O}{N_1/J_1}\right)\left(1 - \frac{\ln Q_O}{N_2}\right) = J_1 J_2$$

9. *Operation of Existing Stagewise Extractor*

For an existing stagewise contactor in which N_{S1} is not equal to N_{S2} the maximum value of $Q_{3/4}$ and optimum values of J_1 and J_2 at fixed $J_1 J_2$ are given by equations 4(2), (6), (10) and (17). The solution is presented graphically in Appendix 41 and a good approximation to the behaviour of these curves is given by

$$\left(\frac{1}{N_1} + \frac{1}{N_2}\right) \ln Q_O = \ln 1/J_1 J_2$$

and

$$\left(1 - \frac{(J_1J_2)^{\mu-1} \ln Q_O}{N_1/J_1}\right)\left(1 - \frac{(J_1J_2)^{\mu} \ln Q_O}{N_2}\right) = 1 - (J_1J_2)^{\mu} \ln 1/J_1J_2$$

where

$$\mu = 0{\cdot}75 \left(\frac{\ln Q_O}{N_2 \ln 1/J_1J_2}\right)^2$$

These approximations become more accurate as J_1J_2 approaches unity and Q_O becomes large.

7.6. Forward and Back Extraction Using Cross-current Flow with Recirculation

1. *Recommended Formula*

For cross-current extraction with recirculation the number of pairs of stages is given by

$$N_R \ln 1/\beta = \ln Q_{3/4}$$

where

$$\beta = \frac{J_2 + J_1J_2}{1 + J_2}$$

2. *Alternative Formulae*

This formula is identical with that given in Conclusion 7.3(*5*) if $\ln 1/J$ is replaced by $\ln 1/\beta$. Any other formula in Table 2.3 may be used if J is also replaced by J_1J_2 in that part of the expression containing Q. For example,

$$N_R \ln 1/\beta = \ln [(1 - J_1J_2)\, Q_{2/4} + J_1J_2]$$

when $J_1J_2 = 1$ this becomes

$$N_R = (1 + J_2)(Q_{2/4} - 1)$$

3. *Minimum Number of Stages*

For a given separation, Q_R and overall operating conditions, J_1J_2 ($= m_1m_2G/L$) the minimum number of cross-current extraction stages occurs when the recirculation rate of the intermediate solvent is infinite, and is given by

$$N_R \ln 1/J_1 J_2 = \ln Q_R$$

This number of stages is always less than the minimum number of forward and back extraction stages N_{SO}, for a given $Q_R = Q_O$ and $J_1 J_2$.

4. Comparison with Counter-current Forward and Back Extraction

In forward and back extraction for a given separation, $Q_R = Q_O$ and overall operating conditions $J_1 J_2$ the minimum number of counter-current stages N_{SO} is always greater than the number of cross-current stages N_R (even when both are calculated at the optimum values of J_1 and J_2 for counter-current flow). Similarly when $N_R = N_{SO}$ the separation Q_O is always less than Q_R for a given $J_1 J_2$.

5. Comparison with Ordinary Counter-current Extraction

(a) CONSTANT $J = J_1 = J_2$

Under the same condition of individual extraction, $J = J_1 = J_2$ ordinary counter-current extraction and cross-current forward and back extraction give the same separation $Q = Q_R$ for a given number of stages $N_S = N_R$.

(b) CONSTANT $J = J_1 J_2$

Under the same conditions of overall extraction, $J = J_1 J_2$ cross-current forward and back extraction always requires more stages than ordinary counter-current extraction for a given separation (or gives a poorer separation from a given number of stages).

7.7. Linear Equilibrium Data

1. Systems Investigated

The solute distribution has been investigated in ninety-nine aqueous–organic systems, forty-seven organic–organic systems and thirty-nine liquid–gas systems as listed in Appendix 6g. A least-squares procedure is used to obtain the linear functional relationship between the concentrations of the two phases.

2. *Tabulation of Slopes*

Systems with an error in the full range slope of less than 1% are listed in Table 6.1. The slopes of the best straight lines passing either through the origin or through the mean values of the observations are given in Table 6.3. Data for systems having a minimum error in the slope of less than 4% are given in Appendices 6h and 6i.

3. *Systems with Error in Slope of less than 1%*

In 38% of the aqueous–organic systems, 26% of the organic–organic systems and 77% of the liquid–gas systems there exists a region with an error in the slope of less than 1%. When the full concentration range is considered the corresponding figures are 16%, 2% and 31%.

4. *Systems with Error in Slope of less than 2%*

In 60% of the aqueous–organic systems, 51% of the organic–organic systems and 80% of the liquid–gas systems there exists a linear region with an error in the slope of less than 2%. When the full concentration range is considered the corresponding figures are 28%, 15% and 46%.

Notation for Chapter 7

This is given at the beginning of the book on page xix and at the end of the individual chapters.

CHAPTER 8

Further Applications

THIS chapter discusses how the results of previous chapters may be extended to more complicated cases and applied to other fields.

GENERAL APPLICATIONS

8.1. Form of Presentation

There are many relationships in chemical engineering in which the final form of the expression is to some extent arbitrary. By considering all the alternative forms in the manner of Chapter 2 it may be that these could be re-expressed in much simpler ways. Any equation involving a separation factor, such as that due to Smoker,[1] or one of the expressions for stagewise and differential back-mixing derived by Hartland and Mecklenburgh,[2] is obviously susceptible to such analysis. It may be also possible to simplify other types of equation by careful reappraisal.

There is much scope in graphical presentation. An equation containing only three variables may be plotted in at least twelve different ways, for there are three choices for the constant parameter and square, log/log or semilog (either axis) graph paper may be used. Comparison of the original plot of Kremser[3] relating Q, J and N_S with Fig. 2.1 shows the improvement which may be effected.

8.2. Errors

Similarly there are many equations which could be analysed in the manner of Chapter 3 to find the effect a variation in one variable has on another.

8.3. Optimisation

Any process which has two balancing factors such as the effect of recirculation rate on the extent of forward and back extraction may be optimised in the manner of Chapter 4.

FORWARD AND BACK EXTRACTION

8.4. Effect of Solvent Washing

The analysis of forward and back extraction in Chapter 4 may be extended to include the effect of washing the intermediate solvent before contactor 2 to remove unwanted solutes extracted in contactor 1. Usually more stages or transfer units are added above the feed in contactor 1 and it becomes a centre-fed contactor. A typical example occurs in the separation of uranium or plutonium from fission products in nuclear fuel reprocessing.[4,5,6]

For a stagewise process the more extraction stages in contactor 1 the more unwanted solute is extracted and hence the more washing stages are needed. In order to optimise the total number of forward and back extraction and washing stages it will therefore be best to err on the side of too few forward extraction stages. A similar argument may be applied to a differential process. The optimum values of N_1 and N_2 worked out for forward and back extraction alone will thus be modified so that N_1 becomes smaller and N_2 larger.

Klinkenberg has discussed the extraction of a single solute in a centre fed extractor[7] and of two solutes in a counter-current extractor.[8,9] Smith and Brinkly[10] have discussed the general case of extraction of a single solute in a centre fed extractor with reflux at both ends. Bush and Denson[11] consider the principles of systematic multiple fractional extraction and show that the greatest fractional separation of two solutes occurs when the volume ratio of the two solvents is inversely equal to the square root of the product of the partition coefficients.

8.5. Minimum Capital Cost

Forward and back extraction could be optimised with respect to other relevant variables.

The length of a differential contactor is proportional to the number of transfer units and the cross-sectional area to the total flowrate. Thus minimising the sum of the products $N_1 (L + S) + N_2 (S + G)$ for a given separation Q_o would minimise the total volume of the system and hence its capital cost.

8.6. Economic Optimisation

As Colburn[12] has pointed out there is an optimum value of J_1 which minimises the sum of the capital and operating costs in absorption. A typical example worked out in detail by Spurlock[13] for the recovery of acetone gave an optimum value of $J_1 = 0{\cdot}7$. For such types of absorption process where the solute is quite valuable, experience showed that the practical operating range of J_1 was usually between 0·5 and 0·8. However, when the solute is less valuable such as in the absorption of refinery gases the operating range of J_1 runs higher, as shown by Sherwood.[14] In stripping, analogously the value of J_2 was ordinarily between 1·25 and 2. Thirty years have elapsed since these rough rules of thumb were formulated and it may be that a careful reappraisal of current operating costs will reveal a shift in the economically optimum values of J_1 and J_2.

It may be possible to obtain a simple relation between the optimum values of J_1 and J_2 which give the minimum number of counter-current forward and back extraction stages for a given separation and the values of J_1 and J_2 which are the most attractive economically. If so the need for iterative complex economic optimisation would be avoided. Figure 4.5 indicates that the optimum values of J_1 and J_2 depend strongly on the overall operating conditions $J_1 J_2$ and on the overall separation Q_o to some extent. Guidance as to the economically optimum value of $J_1 J_2$ would thus be useful.

8.7. Empirical Correlations

Gilliland[15] has devised a method for estimating the number of theoretical plates required for a given separation as a function of the reflux ratio in distillation. A knowledge of the minimum number of plates at total reflux (given by the Fenske[16] equation), and the minimum reflux ratio is required. The correlation was based on plate to plate calculations for several different systems for which typical operating data were available.

By performing similar calculations for the typical operation of forward and back extraction systems it may be possible to devise a correlation giving the actual number of stages for a given recirculation rate in terms of the minimum number of stages or transfer units and the optimum recirculation rate.

A similar correlation could be obtained for cross-current forward and back extraction based on a knowledge of the minimum number of stages required at infinite recirculation. This is given by equation 5(18), which is in some ways analogous to the Fenske[16] equation.

8.8. Curved Equilibrium and Operating Lines

When the equilibrium lines are curved and not straight it should be possible to devise a graphical method for obtaining the optimum solution to forward and back extraction. This may well be based on the operating diagram in Fig. 4.2.

When the solvent flowrates are not constant throughout the contactors or when the solvents are miscible with each other a step by step method is required.

Graphical and step by step methods for obtaining the number of pairs of stages in cross-current extraction with recirculation would be useful.

8.9. Non-equilibrium Stages

When the stages are not at equilibrium the mass transfer from the x to y phase in the nth stage of volume V is $k_G aV(mx_n - y_n)$. A solute balance round the nth stage, as in Appendices 1a and 2b, separating the variables in the manner of Appendix 1b leads to a finite difference equation which may be solved using the boundary conditions A1a(2)–(5). This gives expressions for the concentration differences listed in Appendix 2b which show that the number of non-equilibrium stages, N_A, is given by

$$N_A \ln 1/\alpha = \ln Q \tag{8(1)}$$

where

$$Q = \frac{y_{\text{in}} - y(x_{\text{out}})}{y_{\text{out}} - y(x_{\text{in}})} \tag{8(2)}$$

$$\alpha = \frac{1 + Jt}{1 + t} \tag{8(3)}$$

$$t = k_G aV/G \tag{8(4)}$$

and

$$J = mG/L \tag{8(5)}$$

The equation may be inverted if N_A, Q, α, J and t are replaced by their inverses N_A, $1/Q$, $1/\alpha$, $1/J$ and Jt respectively. Alternatively we may introduce the Murphree stage efficiency based on the concentration y defined by

$$E_m = \frac{y_n - y_{n-1}}{y(x_n) - y_{n-1}} \tag{8(6)}$$

which expresses the closeness of approach to equilibrium in the nth stage. This is related to the transfer coefficient t by

$$E_m = \frac{t}{1 + t} \tag{8(7)}$$

so that

$$\alpha = 1 + (J - 1) E_m \tag{8(8)}$$

When the stages approach equilibrium, t tends to infinity, E_m approaches unity and α approaches J.

If α is replaced by J and N_A by N_S, equation 8(1) becomes identical with formula VII of Table 2.3, which applies to equilibrium stages, and so with these modifications Fig. 2.1 may be used for non-equilibrium calculations. Q is related by the overall balance involving J to the other separation factors listed in Table 2.3. For example, Marshall and Pigford[17] have derived

$$N_A \ln 1/\alpha = \ln [(1 - J) Q + J] \tag{8(9)}$$

where

$$Q = \frac{y_{in} - y(x_{in})}{y_{out} - y(x_{in})} \tag{8(10)}$$

which is analogous to formula V of Table 2.3 derived by Kremser[3]. Hartland and Mecklenburgh[2] have derived

$$N_A \ln 1/\alpha^* = \ln \frac{1 - J^* Q^*}{1 - Q^*} \tag{8(11)}$$

where

$$Q^* = \frac{y(x_{out}) - y(x_{in})}{y_{in} - y(x_{in})} \tag{8(12)}$$

which is analogous to the inverse of formula I in Table 2.3 derived by Souders and Brown[18].

It would be useful to derive equivalent formulae for counter-current and cross-current forward and back extraction when the stages are not at equilibrium.

8.10. Cross-current Extraction

The analysis of cross-current extraction in Chapter 5 may be extended to include the effect of recycling the intermediate solvent around more than one pair of stages. The intermediate solvent may flow counter-current to the initial and final solvent for one, two, three or more stages before being recycled to the other contactor. In addition, the analysis may be extended to a differential process, recycling the intermediate solvent around different numbers of transfer units in contactors 1 and 2.

LINEAR EQUILIBRIUM DATA

8.11. Effect of Concentration Units

As pointed out in Chapter 6 the linearity of a system depends to some extent upon the concentration units used, as density is a function of concentration.[19] It is possible for a system to be non-linear on the basis of g/l units say, and yet be linear if instead the units were converted to mole fractions. Also the concentrations of all the liquid–liquid systems described have been expressed on the basis of the entire phase (i.e. taking into account all the three components). It is, however, also permissible to express the concentration on a consolute-free basis, as Ishida[20] has done, and this form is sometimes more useful in carrying out solvent extraction calculations. Many permutations of the units are possible, and it is possible that for any given system one of these could give rise to a linear plot.

8.12. Effect of Concentration Range

In Chapter 6 straight lines have been fitted in the upper and lower concentration regions of the distribution curve. It is probable that many systems will also have a range in the middle of the curve which is linear. However, it is even more difficult to tabulate such regions concisely as both pairs of end concentrations can then vary simultaneously.

Notation for Chapter 8

The notation for forward and back extraction is given in Chapter 4.

a interfacial area per unit volume
c intercept of equilibrium line on y-axis
E_m Murphree stage efficiency based on concentration y ($E_m = t/(1 + t)$)
G flowrate of phase in which solute concentration is y
J extraction factor ($J = mG/L$)

k_G	overall mass transfer coefficient based on phase G
L	flowrate of phase in which solute concentration is x
m	slope of equilibrium line (equation $y = mx + c$)
n	stage number
N	number of non-equilibrium stages
Q	separation factor
t	stagewise transfer coefficient based on phase G ($t = k_G aV/G$)
V	volume of a stage
x	solute concentration in phase of flowrate L
$x(y)$	x concentration in equilibrium with concentration y ($x(y) = (y - c)/m$)
y	solute concentration in phase of flowrate G
$y(x)$	y concentration in equilibrium with concentration x

Greek symbols

α	extraction factor for non-equilibrium stages ($\alpha = (1 + Jt)/(1 + t)$)

Subscripts

in	refers to stream entering contactor
out	refers to stream leaving contactor
A	refers to actual stages
G	refers to phase of flowrate G
n	refers to nth stage
$n - 1$	refers to $(n - 1)$th stage

Superscript

*	denotes inversion

The above quantities may be expressed in any set of consistent units in which force and mass are not defined independently.

References

1. E. H. SMOKER, Analytic determination of plates in fractionating columns, *Am. Inst. Chem. Engrs.* **34**, 163 (1938).
2. S. HARTLAND and J. C. MECKLENBURGH, Comparison of stagewise and differential two phase counter-current extraction with backmixing, *Chem. Eng. Sci.* **21**, 1209 (1966).

3. A. KREMSER, Theoretical analysis of absorption process, *Nat. Pet. News* **22** (21), 42 (1930).
4. F. R. BRUCE *et al.* (Editors), *Progress in Nuclear Energy*, Series III. *Process Chemistry*, Vol. 1, Chap. 5, Wet process for radiochemical separations, McGraw-Hill Book Co. Inc., New York, 1956.
5. D. G. KIRRAKER, Temperature effects on TBP solvent extraction processes, *Proceedings of Second United Nations International Conference on the Peaceful Uses of Atomic Energy*, Vol. 17, p. 333, United Nations, Geneva, 1958.
6. J. R. FLANARY, Solvent extraction separation of uranium and plutonium from fission products by means of tributyl phosphate, *Proceedings of the International Conference on the Peaceful Uses of Atomic Energy*, Vol. 9, p. 528, United Nations, New York, 1956.
7. A. KLINKENBERG, Calculation of the efficiency of counter-current stagewise mass transfer processes with constant distribution factor, when in the stationary state. I. Distribution of one component only, *Chem. Eng. Sci.* **1**, 86 (1951).
8. A. KLINKENBERG, R. A. LAUWERIER and G. H. REMAN, Calculation of the efficiency of counter-current stagewise mass transfer processes with constant distribution factor, when in the stationary state. II. Stage requirements for the separation of two components by two solvent extraction. *Chem. Eng. Sci.* **1**, 93 (1951).
9. A. KLINKENBERG, Two component separation by double solvent extraction, *Ind. Eng. Chem.* **45**, 653 (1953).
10. B. D. SMITH and W. K. BRINKLEY, General short cut equation for equilibrium stage processes, *Am. Inst. Chem. Engrs.* **6** (3), 446 (1960).
11. M. T. BUSH and P. M. DENSON, Systematic multiple fractional extraction procedures. Principles of application to separation of organic mixtures, *Anal. Chem.* **20**, 121 (1948).
12. A. P. COLBURN, The simplified calculation of diffusional processes. General consideration of two-film resistances, *Trans. Am. Inst. Chem. Engrs.* **35**, 211 (1939).
13. B. SPURLOCK, Estimate of costs for construction and operation of a solvent recovery system, *Trans. Am. Inst. Chem. Engrs.* **31**, 575 (1935).
14. T. K. SHERWOOD, *Absorption and Extraction*, p. 110, McGraw-Hill Book Co. Inc., New York, 1937.
15. E. R. GILLILAND, Multicomponent rectification—estimation of the number of theoretical plates as a function of the reflux ratio, *Ind. Eng. Chem.* **32**, 1220 (1940).
16. M. R. FENSKE, Fractionation of straight run Pennsylvania gasolene, *Ind. Eng. Chem.* **24**, 482 (1932).
17. W. R. MARSHALL and R. L. PIGFORD, *The Application of Differential Equations to Chemical Engineering Problems*, p. 80, University of Delaware, Newark, Delaware, 1947.
18. M. SOUDERS and G. C. BROWN, Fundamental design of absorbing and stripping columns for complex vapours, *Ind. Eng. Chem.* **24**, 519 (1932).
19. J. Y. OLDSHUE and J. H. RUSHTON, Continuous extraction in a multistage mixer column, *Chem. Eng. Progress* **48**, 297 (1952).
20. K. ISHIDA, Solvent selectivity of liquid ammonia for pure hydrocarbons—tie line data and comparison of selectivity with other solvents, *Bull. Chem. Soc. Japan* **30**, 612 (1957).

Index